Compostable Polymer Materials

Compostable Polymer Materials

Ewa Rudnik

Amsterdam • Boston • Heidelberg • London • New York • Oxford
Paris • San Diego • San Francisco • Singapore • Sydney • Tokyo

Elsevier
The Boulevard, Langford Lane, Kidlington, Oxford OX5 1GB, UK
Radarweg 29, PO Box 211, 1000 AE Amsterdam, The Netherlands

First edition 2008

British Library Cataloguing-in-Publication Data
A catalogue record for this book is available from the British Library

Library of Congress Cataloging-in-Publication Data
A catalog record for this book is available from the Library of Congress

ISBN: 978-0-08-045371-2

For information on all elsevier publications
visit our website at www.books.elsevier.com

Typeset by Charon Tec Ltd (A Macmillan Company), Chennai, India
www.charontec.com

Printed and bound in Hungary

08 09 10 10 9 8 7 6 5 4 3 2 1

Contents

Preface

Let us permit nature to have her way. She understands her business better than we do.
Michel do Montaigne
If one way be better than another, that you may be sure is Nature's way.
Aristotle

Composting is Nature's way of recycling. When a plant dies, the stem, flowers and leaves are broken down by microbes in the soil and the nutrients they contain will feed the seed to grow into a seedling and then into a mature plant. The cycle is complete. The natural decay processes that occur in soil include raw organic materials like manure, leaves, grass clippings, food wastes and municipal biosolids. During composting organic residues are converted to stable soil-like humic substances, called compost. This final product is a valuable soil resource for agricultural, horticultural and silvicultural purposes.

Composting is probably the oldest recycling technique in the world. The art of composting dates back to the early Greeks and Romans. There are also biblical references to composting. The Chinese are thought to be the first people to develop larger composting sites for use in farming. George Washington, the first president of the USA, was the first recognized composter in America. The knowledge and practices of composting were passed from country to country, down through generations until today where modern composting facilities have been developed.

The composting process used in municipal or industrial composting facilities is only mimicking and speeding up what nature is doing every day. The difference is that the degradative rates in composting are higher because of the higher temperatures.

The composting industry has expanded significantly in recent years and this trend seems set to continue in different regions in the world, e.g. in Europe, America and in Asian countries like China. Environmental awareness has led to a growing interest in developing an environmentally friendly manner of disposal of municipal and industrial wastes.

It is known that plastics waste contribute to a large volume of municipal waste. In general, waste management strategy puts a growing emphasis on the three Rs: Reduce, Reuse, and Recycle. Composting, recognized as organic recycling, provides a means of accomplishing these objectives. It is noteworthy that composting as a technology is adaptable and suitable for treating wastes in a variety of socioeconomic and geographical locations.

The aim of the book is to describe in a coherent manner the special class of polymers especially designed to be disposed of after their useful life in composting facilities. Compostable plastics are polymers that undergo degradation by biological processes during composting to yield CO_2, water, inorganic compounds and biomass at a rate consistent with known compostable materials (e.g. cellulose).

The book is concerned with the hot topic of dealing with a family of polymers that are designed to be degraded in industrial and municipal compost facilities.

Recently, compostable packaging materials were introduced into the market to reduce the amounts of conventional packaging materials and at the same time be recovered by the municipal organic waste collection system. An example of commercial implementation of compostable polymers includes Cargill Dow Nature Works polylactic acid (PLA), which is produced at a rate of 140 000 tonnes/year at Blair, Nebraska (USA), via carbohydrate fermentation. Other global companies are also producing compostable polymer materials. NEC is using a PLA composite with kenaf fibres for laptop computer cases. Polyhydroxyalkanoates produced by microbial "biofactories" are suitable for films, fibres, adhesives, coatings, moulded goods and a variety of other applications.

The largest and most successful effort to date to recycle organic waste and use compostable cutlery and food serviceware was at the 2000 Olympic Games held in Sydney (Australia) where the collection and composting of the wastes generated by the nine million visitors to the Games resulted in the recovery of 76% of the solid wastes generated.

I hope that this book may be useful in developing ideas about composting and provides information for researchers and students interested in materials science as well as ecological issues. The book covers the entire spectrum of preparation, properties, degradation, and environmental issues of this kind of polymer. Emphasis is given to the recent studies concerning compostability and ecotoxicological assessment of polymer materials – important issues from an ecological point of view. Moreover, thermal behaviour of compostable polymers is described. Future perspectives, including price evolution during last decade and market estimation, are presented.

Chapter 1 introduces the problem, explaining the role of compostable polymer materials in a sustainable development and the reasons for the growing interest.

Chapter 2 gives an overview of compostable polymer materials starting with definitions, and explaining differences in comparison with biodegradable polymers. It discusses origin (synthetic and natural), structures and methods of preparation of compostable polymers. The producers of compostable polymer materials and their websites are given.

Chapter 3 summarizes the main properties of compostable polymers, i.e. physico-chemical, mechanical, and thermal properties. It also contains data about their processing and current applications.

Chapter 4 describes the behaviour of compostable polymer materials during degradation in different environments (inert and oxidative atmospheres). An overview concerning thermal stability and apparent activation energy of decomposition is presented.

Chapter 5 gives an overview of the composting process and methods, including up-to-date standardized guidelines for evaluating compostability of polymer materials. It also contains information about compost quality standards as well as the description of certification systems for compostability used in different regions in the world (Europe, USA, Japan).

Chapter 6 summarizes current testing methods used for the biodegradability testing of compostable polymer materials, standardized as well as non-standardized, reported in the literature. The focus is on studies under composting conditions.

Chapter 7 gives the definitions related to ecotoxicity testing, describes the currently used method and gives an overview of ecotoxicological assessment of compostable polymer materials.

Chapter 8 describes the environmental impact of compostable polymer materials, including life cycle analysis (LCA).

Chapter 9 presents the future prospects for compostable polymer materials. The price evolution during the last decade, manufacturing capacity, recent legislative measures as well as potential markets are presented.

The book contains references, an index, as well as symbols and acronyms used in the text.

It is difficult to acknowledge everyone who has helped me in the preparation of this book. Many lectures and publications were stimulating in developing ideas regarding the creation of a special book concerning compostable polymer materials. Let me list among them Professors

Raman Narayan and Anne Christine Albertsson, who promote the ideas for and share the passion about environmentally friendly polymers. My sincere thanks also go to my editor Derek for his outstanding patience.

I would like to thank most warmly my friends from Industrial Chemistry Research Institute and elsewhere for their support and friendship. I am fortunate to have met in my professional career many persons who have showed me kindness and given their help. Thank you very much.

Special thanks are also due to my beloved family: my parents and my brothers for their love and faith in me. This monograph is dedicated to you.

Ewa Rudnik

Symbols and abbreviations

AAC – aliphatic–aromatic copolyesters
APME – Association of Plastic Manufacturers in Europe
ASTM – American Society for Testing Materials
BOD – biochemical oxygen demand
BPI – The Biodegradable Products Institute (USA)
BPS – Biodegradable Plastics Society (Japan)
CA – cellulose acetate
CAB – cellulose acetate butyrate
CAP – cellulose acetate propionate
CEN – European Organization for Standardization
DIC – dissolved inorganic carbon
DIN – German Organization for Standardization
DMT – trimethyl terephthalate
DOC – dissolved organic carbon
EC – effective concentration
ED – effective dose
GHG – greenhouse gas
IC – inorganic carbon
ISO – International Organization for Standardization
JIS – Japanese Standards Association
LC – lethal concentration
LCA – life cycle assessment
LD – lethal dose
LOEC – lowest observed effect concentration
MSW – municipal solid waste
NOEC – no observed effect concentration
OECD – Organization for Economic Cooperation and Development
PBAT – poly(butylene adipate terephthalate)
PBS – poly(butylene succinate)
PBSA – poly(butylene succinate adipate)
PBST – poly(butylene succinate terephthalate)
PBT – poly(butylene terephthalate)
PCL – poly(ϵ-caprolactone)
PDLA – poly(D-lactide)
PDO – 1,3-propanediol
PEA – polyesteramides
PES – poly(ethylene succinate)
PESA – poly(ethylene succinate adipate)
PET – poly(ethylene terephthalate)
PHA – polyhydroxyalkanoates
PHB – poly(3-hydroxybutyrate)
PHBV – poly(3-hydroxybutyrate-*co*-3-hydroxyvalerate)
PHH – poly(hydroxyhexanoate)

PHV – poly(3-hydroxyvalerate)
PLA – poly(lactic acid) or polylactide
PLLA – poly(L-lactide)
PTA – purified terephthalic acid
PTMAT – poly(tetramethylene adipate terephthalate)
PTT – poly(trimethylene terephthalate)
PVA – poly(vinyl alcohol)
PVAc – poly(vinyl acetate)
ROP – ring-opening polymerization
Thbiogas – theoretical amount of evolved biogas
ThCH$_4$ – theoretical amount of evolved methane
ThCO$_2$ – theoretical amount of evolved carbon dioxide
ThOD – theoretical oxygen demand
TOC – total organic carbon
TPS – thermoplastic starch

Chapter 1
Introduction

Chapter 1
Introduction

Polymer materials with a range of excellent mechanical properties, low density, durability and low cost, are widely used in the daily needs of contemporary society, ranging from simple packaging to heavy construction, and play important role in the improvement and quality of life. However, due to their persistence in the environment, polymer materials present a danger to our ecosystems.

With economic growth resulting in an increase in the amount of waste generated over the past decades, plastic waste becomes an environmental problem of growing concern in the world. It is expected that during the beginning of the 21st century there will be two- to three-fold increases in plastics consumption, particularly due to growth in developing countries [1]. However, this increased use of plastics is accompanied by a rapid accumulation of solid waste and plastics litter, which, due to their resistance to biodegradation, have a deleterious effect on the environment as an obvious contributor to pollution.

The worldwide increase in plastics waste has involved, within the global vision of environmental protection and sustainability, a great deal of action and strategies aimed at minimizing the negative impact of the increasing production and consumption of polymer materials.

In general, waste strategies employed in different regions in the world are similar and are based on the prevention and recycling of waste. For example, Japan has extensive legislation related to waste and other sustainable production and consumption policies under the "3R-reducing, reusing and recycling" umbrella.

The strategy of the EU to cope with waste is to:

- prevent waste in the first place;
- recycle waste;
- optimize the final disposal of waste.

In response to the growing challenges of waste production and management, the European Parliament and the Council have adopted a number of Directives to ensure that waste is recovered or disposed of without impairing the environment and human health.

According to the European Directive on Packaging and Packaging Waste [2] the management of packaging and packaging waste should include as a first priority the prevention of packaging waste and, as additional fundamental principles, reuse of packaging, recycling and other forms of recovering packaging waste and, hence, reduction of the final disposal of such waste.

Prevention means the reduction of the quantity and harmfulness to the environment of:

- materials and substances contained in packaging and packaging waste;
- packaging and packaging waste at production process level and at marketing, distribution, utilization and elimination stages, in particular by developing "clean" production methods and technology.

Reuse is defined as any operation by which packaging, which has been conceived and designed to accomplish within its life cycle a minimum number of trips or rotations, is refilled

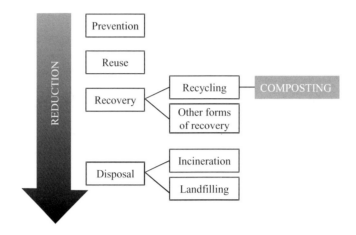

Figure 1-1 Plastics waste treatment strategy.

or used for the same purpose for which it was conceived, with or without the support of the auxiliary products present on the market enabling the packaging to be refilled.

Recovery includes operations provided for in Annex II.B to Directive 75/442/EEC on waste [2], e.g. use as a fuel or other means to generate energy, recycling/reclamation of organic substances which are not used as solvents (including composting and other biological transformation processes).

Energy recovery means the use of combustible packaging waste as a means to generate energy through direct incineration with or without other waste but with recovery of the heat.

Recycling is defined as the reprocessing in a production process of the waste materials for the original purpose or for other purposes including organic recycling but excluding energy recovery.

Disposal operations include deposit into or onto land (e.g. landfilling), incineration, etc.

The use of compostable plastics is one valuable recovery option (biological or organic recycling). According to the EU Directive on Packaging and Packaging Waste [1] **organic recycling** means the aerobic (**composting**) or anaerobic (biomethanization) treatment, under controlled conditions and using microorganisms, of the biodegradable parts of packaging waste, which produces stabilized organic residues or methane. Landfill is not considered as a form of organic recycling.

The Waste Management Hierarchy, i.e. minimization, recovery and transformation, and land disposal have been adopted by most developed countries with strategies used depending on such factors as population density, transportation infrastructure, socioeconomic and environmental regulations.

1.1. SITUATION IN EUROPE

Current EU waste policy is based on a concept known as the waste hierarchy. This means that, ideally, waste should be prevented and what cannot be prevented should be reused, recycled and recovered as much as is feasible, with landfill being used as little as possible. The long-term goal is for the EU to become a recycling society.

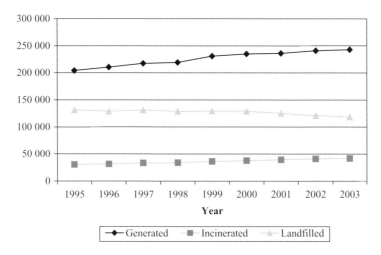

Figure 1-2 Municipal waste generated, landfilled and incinerated by EU countries (25) from 1995 to 2003 in 1000 t.

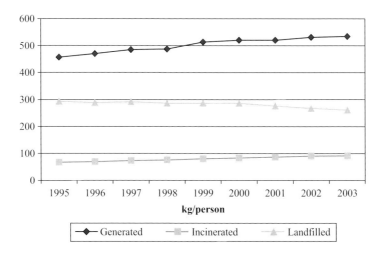

Figure 1-3 Municipal waste generated, landfilled and incinerated by EU countries (25) per capita from 1995 to 2003 in kg/person.

Despite the intensive efforts of some countries to reduce the amounts of waste, the quantity of solid waste is significantly increasing within the European Union.

From 1995 to 2003 municipal waste generation in the European Union (EU 25) has constantly grown by about 2% per year from 204 million tonnes (457 kg/person) in 1995 to 243 million tonnes (534 kg/person) in 2003 (cf. Fig. 1.2) [4].

Where municipal waste is concerned, each EU citizen produces an average of 550 kg per year (Fig. 1.3). Generation is higher in the old Member States with 574 kg/person compared to 312 kg/person in the new Member States.

The landfilled waste has decreased in the same period by about 10% from 131.4 million tonnes in 1995 to 118.5 million tonnes in 2003 on account of increased incineration and

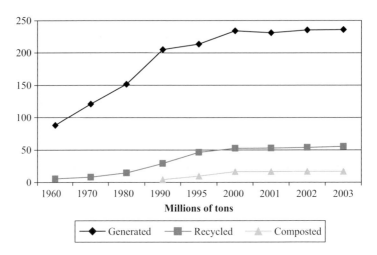

Figure 1-4 Municipal waste generated, recovered for recycling and composting in the USA from 1960 to 2003 in millions of tons.

recycling rates. In 2003, 48.8% of the municipal waste generated was landfilled, 17.3% was incinerated and 33.9% was recycled or treated otherwise. Recycling has gained an important role in nearly all 15 EU (old) countries, and accounts for the treatment of up to 33% (Germany) of the municipal waste total in 2002. Composting contributes considerably to waste management in several countries such as Belgium, Denmark, Germany, Spain, France, Italy and the Netherlands. Between 13% and 24% of municipal waste is treated by composting in these countries, the composted amounts ranging between 71 kg/person in France and 147 kg/person in the Netherlands.

1.2. SITUATION IN THE UNITED STATES

Since 1980, the total annual generation of municipal solid waste has increased by more than 50% to its 2003 level, i.e. 236.2 million tons per year [5].

Organic materials are the largest components of MSW in the USA. Paper and paperboard products account for 35% of the waste stream, followed by yard trimmings and food scraps with about 24% (2003). Plastics comprise 11%, i.e. 26.7 millions tons, at third place in municipal solid waste composition. It is noteworthy that containers and packaging made up the largest portion of waste generated, about 75 million tons. Nearly 9% of plastic containers and packaging was recycled, compared with 22% of glass containers and 15% of wood packaging recovered for recycling.

1.3. SITUATION IN OTHER REGIONS OF THE WORLD

The significant increase in plastics consumption is also observed in other regions of the world. For example, rapid industrialization and economic development in Singapore have caused a tremendous increase in solid waste generation. The annual amount of disposed solid waste increased from 0.74 million tonnes in 1972 to 2.80 million tonnes in 2000 [6]. It is estimated that solid waste generation in Singapore has amounted to about 4.5–4.8 million tonnes per year. Plastics account for 5.8% of the total solid waste, in third position after food waste (38.3%) and paper/cardboard (20.60%). Taking into account that plastic bags and bottles have become one of the major solid waste streams, using waste plastics to manufacture polymer concrete and developing biodegradable plastics has received much attention in recent years.

In Australia, the annual plastics consumption has increased from 1 336 386 in 1997 to 1 521 394 tonnes in 2003, whereas the total recycling rate of plastics has increased from 7.0% to 12.4% [7]. It is noteworthy that plastics packaging recycling in 2003 was 134 905 tonnes, which is 20.5% of packaging consumption during a year.

In China, the production of municipal solid waste (MSW) and sewage sludge is changing rapidly along with economic development [8]. The amount of solid waste produced in China is large and is increasing rapidly. The average amount of MSW produced by each person daily increased from 1.12 to 1.59 kg from 1986 to 1995 [8]. In 1995, China produced 644.74 million tonnes of industrial solid waste and 237 million tonnes of MSW. At present it is estimated that the amount of municipal refuse produced by each person annually is about 204.4–440 kg, and the total solid waste produced in China is about 27.15% of that in Asia and 15.07% of that globally [8]. About 85% of the total amount of MSW production in China is in cities, e.g. about 60% of the total amount of MSW produced in China is in 52 cities, whose population is over 0.5 billion. MSW in China is mainly treated by landfilling and composting, and a smaller amount of MSW is treated by incineration. The amount of MSW treated by landfilling and composting is over 70% and 20% of the total amount of MSW disposed of, respectively [8]. Composting has emerged as a potentially viable alternative by local governments because of lower investment and operation costs.

However, despite the efforts that have been made, overall waste volumes are growing. Management of plastics waste remains a problem. The observed increased plastics consumption throughout the world makes the development of more recyclable and/or biodegradable plastics necessary to reduce the amount of plastics to landfill. According to an amendment to the European Directive on Packaging and Packaging Waste [9] recovery and recycling of packaging waste should be further increased to reduce its environmental impact.

Compostable polymers, which have been designed to be disposed after their useful life by means of organic recycling, i.e. composting, are one of the strategic options available for the management of plastics waste. Composting is an attractive alternative for reducing solid waste and is especially suitable for those segments of conventional plastics in which recycling is difficult or economically not feasible.

The growing environmental awareness and new rules and regulations, as well as new trends in solid waste management, have led scientists to increase activities on the design of compostable polymer materials that easily degrade under well-defined environmental conditions.

REFERENCES

[1] Workshop "Promotion of Sustainable Plastics" 29–30 November 2005, San José, Costa Rica.
[2] European Parliament and Council Directive 94/62/EC of 20 December 1994 on packaging and packaging waste.
[3] Council Directive 75/442/EEC on waste, as amended by Council Directive 91/156/EEC.

[4] Waste generated and treated in Europe. Data 1995–2003, European Commission, Eurostat, Luxembourg, 2005.

[5] Municipal Solid Waste Generation, Recycling, and Disposal in the United States: Facts and Figures for 2003, United States Environmental Protection Agency.

[6] Bai R., Sutanto M.: The practice and challenges of solid waste management in Singapore, *Waste Management* 22 (2002) 557–567.

[7] National Recycling Survey 2004, Main Survey Report, PACIA Nolan ITU, September 2004.

[8] Wei Y.-S., Fan Y.-B., Wang M.-J., Wang J.-S.: Composting and compost application in China, *Resourc. Conserv. Recycl.* 30 (2000) 277.

[9] Directive 2004/12/EC of the European Parliament and Council of 11 February 2004 amending Directive 94/62/EC of 20 December 1994 on packaging and packaging waste.

Chapter 2
Compostable polymer materials – definitions, structures and methods of preparation

Chapter 2
Compostable polymer materials – definitions, structures and methods of preparation

"Biodegradable polymers" or "compostable polymers" were first commercially introduced in the 1980s. These first-generation biodegradable products were made from a conventional polymer, usually polyolefin (e.g. polyethylene) mixed together with starch or some other organic substance. When starch was eaten by microorganisms, the products were broken down, leaving small fragments of polyolefins.

In 1994 Narayan *et al.* wrote: "The U.S. biodegradables industry fumbled at the beginning by introducing starch filled (6–15%) polyolefins as true biodegradable materials. These at best were only biodisintegradable and not completely biodegradable. Data showed that only the surface starch biodegraded, leaving behind a recalcitrant polyethylene material" [1].

The situation confused consumers and government regulators, and put into question the biodegradable plastics market for some years. Since then the confusion or misunderstanding appeared about what was and what was not biodegradable and/or compostable. Additionally, no scientifically based test methods or standards existed to support claims made by plastics manufacturers for the "biodegradability" or "compostability" of their products.

More recently, international and national standards bodies, i.e. International Organization for Standardization (ISO), American Society for Testing and Materials (ASTM), Japanese Standards Association (JIS) and European Organization for Standardization (EN), have developed definitions related to the degradation of plastics. Nowadays, ISO and ASTM standards exist describing in detail the purposes of "biodegradable" and "compostable".

The ASTM D6400 standard establishes the requirements for the labelling of materials and products, including packaging made from plastics, as "compostable in municipal and industrial composting facilities".

Table 2.1. Definitions of compostability according to ASTM D6400 [2]

Compostable plastic
A plastic that undergoes degradation by biological processes during composting to yield carbon dioxide, water, inorganic compounds, and biomass at a rate consistent with other known compostable materials and leaves no visually distinguishable or toxic residues.

Composting
A managed process that controls the biological decomposition and transformation of biodegradable materials into a humus-like substance called compost: the aerobic mesophilic and thermophilic degradation of organic matter to make compost, the transformation of biologically decomposable material through a controlled process of biooxidation that proceeds through mesophilic and thermophilic phases and results in the production of carbon dioxide, water, minerals and stabilized organic matter (compost or humus). Composting uses a natural process to stabilize mixed decomposable organic material recovered from municipal solid waste, yard trimmings, biosolids (digested sewage sludge), certain industrial residues and commercial residues.

Degradable plastic
A plastic designed to undergo a significant change in its chemical structure under specified environmental conditions, resulting in a loss of some properties that may be measured by standard test methods appropriate to the plastic and the application in a period of time that determines its classification.

ISO/DIS 17088 specifies test methods and requirements to determine and label plastic products and products made from plastics that are designed to be recovered through aerobic composting. It particularly establishes the requirements for labelling of materials and products, including packaging made from plastics, as "compostable", "compostable in municipal and industrial composting facilities" and "biodegradable during composting".

The definition of "compostable plastic" proposed in ISO/DIS 17088 is identical to that given in the ASTM D 6400 standard.

Table 2.2. Definitions of compostability according to ISO/DIS 17088 [3]

Compostable plastics
A plastic that undergoes degradation by biological processes during composting to yield CO_2, water, inorganic compounds, and biomass at a rate consistent with other known compostable materials and leaves no visible, distinguishable or toxic residue.

Composting
The autothermic and thermophilic biological decomposition of biowaste (organic waste) in the presence of oxygen and under controlled conditions by the action of micro- and macroorganisms in order to produce compost.

Compost
Organic soil conditioner obtained by biodegradation of a mixture consisting principally of vegetable residues, occasionally with other organic material and having a limited mineral content.

Disintegration
The physical breakdown of a material into very small fragments.

In spite of its very large use (and abuse) term "biodegradable" is not helpful because it is not informative. The term does not convey any information about the specific environment where the biodegradation is supposed to take place, the rate that will regulate the process (fast, slow), and the extent of biodegradation (partial or total conversion into CO_2).

The definition of "biodegradable" has been assessed during the past decade. Some examples of definitions of "biodegradable plastic" are given below.

ASTM definition [2]: "a degradable plastic in which the degradation results from the action of naturally occurring microorganisms such as bacteria, fungi, and algae".

ISO and CEN definition [4]: "degradable plastic in which degradation results in lower molecular weight fragments produced by the action of naturally occurring microorganisms such as bacteria, fungi and algae".

According to ISO definition [4] degradable plastic means "A plastic designed to undergo a significant change in its chemical structure under specific environmental conditions resulting in a loss of some properties that may vary as measured by standard test methods appropriate to the plastic and the application in a period of time that determines its classification."

Japanese Biodegradable Polymers Society (BPS) defines biodegradable plastics (called GreenPla) as plastics which can be used as conventional plastics, while on disposal they decompose to water and carbon dioxide by the action of microorganisms commonly existing in the natural environment [5].

Most of the definitions of biodegradation are based on the same concept: the action of microorganisms on the material and its conversion into carbon dioxide or methane and water.

A plastic can be degradable without being biodegradable, i.e. it might disintegrate into pieces or even an invisible powder, but not be assimilated by microorganisms. A plastic can be degradable and even biodegradable without being compostable, i.e. it might biodegrade at a rate that is too slow to be called compostable [6].

The difference between biodegradable and compostable polymers lies in additional requirements related to the latter. Besides biodegradation into carbon dioxide, water, inorganic compounds, and biomass compostable polymers must fulfil other criteria such as compatibility with the composting process, no negative effect on quality of compost and a degradation rate consistent with other known composting materials.

It is noteworthy that compostable plastics are *a priori* designed for a given method of safe disposal, i.e. composting. This means that after their useful life they will biodegrade in a composting process. The idea of compostable polymers is in agreement with life cycle thinking.

To summarize, the requirements a material must satisfy to be termed "compostable" include mineralization (i.e. biodegradation to carbon dioxide, water and biomass), disintegration into a composting system, and completion of its biodegradation during the end-use of the compost, which, moreover, must meet relevant quality criteria, e.g. no ecotoxicity. The satisfaction of requirements should be proved by standardized test methods. These requirements and test methods are described in detail in Chapters 5 and 6.

Compostable polymers can be divided according to source of origin or method of their preparation (Fig. 2.1).

On the basis of origin, compostable polymers are derived from renewable and petrochemical resources.

Biodegradable polymers from renewable resources include:

1. Polylactide (PLA).
2. Polyhydroxyalkanoates: poly(3-hydroxybutyrate) (PHB).
3. Thermoplastic starch (TPS).
4. Cellulose.
5. Chitosan.
6. Proteins.

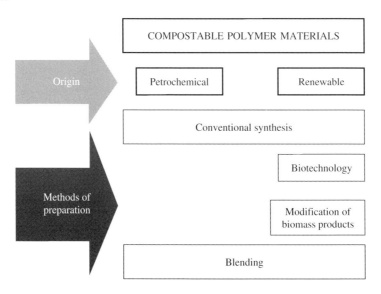

Figure 2-1 Classification of compostable polymers.

Biodegradable polymers from petroleum sources comprise:

1. Aliphatic polyesters and copolyesters (e.g. poly(butylene succinate) – PBS; poly(butylene succinate adipate) – PBSA).
2. Aromatic copolyesters (e.g. poly(butylene adipate terephthalate) – PBAT).
3. Poly(ε-caprolactone) – PCL.
4. Polyesteramides – PEA.
5. Poly(vinyl alcohol) – PVA.

There are three principal ways to produce polymers from renewable resources, i.e. bio-based polymers, i.e.:

1. to make use of natural polymers which may be modified but remain intact to a large extent (e.g. starch polymers);
2. to produce bio-based monomers by fermentation which are then polymerized (e.g. polylactic acid);
3. to produce bio-based polymers directly in microorganisms or in genetically modified crops (polyhydroxyalkanoates).

In general, on the basis of methods of preparation, compostable polymer materials can be prepared via:

1. conventional synthesis:
    *   polymerization from non-renewable monomer feedstocks, e.g. poly(ε-caprolactone) – PCL – copolyesters;
    *   polymerization from renewable monomer feedstocks, e.g. polylactic acid;
2. biotechnological route (extraction, fermentation), e.g. poly(hydroxybutyrate-*co*-hydroxyvalerate) – PHBV;
3. preparation directly from biomass, e.g. plants – starch;
4. blending, e.g. starch–polycaprolactone blends.

A method based on blending of biodegradable polymers is very often used in order to improve the properties of compostable polymer materials or to decrease their cost. The various polymers used are both renewable and of petrochemical origin. Novamont's Mater-Bi is an example of such a material.

2.1. BIODEGRADABLE POLYMERS FROM RENEWABLE RESOURCES

2.1.1 Poly(lactic acid) – PLA

The molecular structure of polylactic acid (PLA) is schematically presented in Fig. 2.2. PLA, linear aliphatic thermoplastic polyester, is prepared from lactic acid. Lactic acid (2-hydroxy propionic acid) is one of the simplest chiral molecules and exists as two stereo isomers, L- and D-lactic acid (Fig. 2.3).

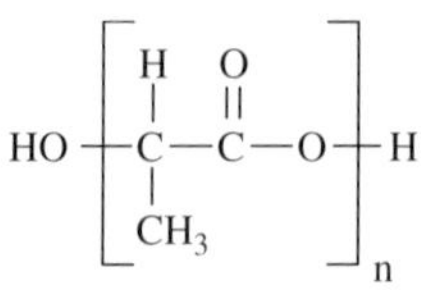

$$HO \left[\begin{array}{c} H \\ | \\ C \\ | \\ CH_3 \end{array} - \begin{array}{c} O \\ || \\ C \end{array} - O \right]_n H$$

Figure 2-2 Structure of poly(lactic acid).

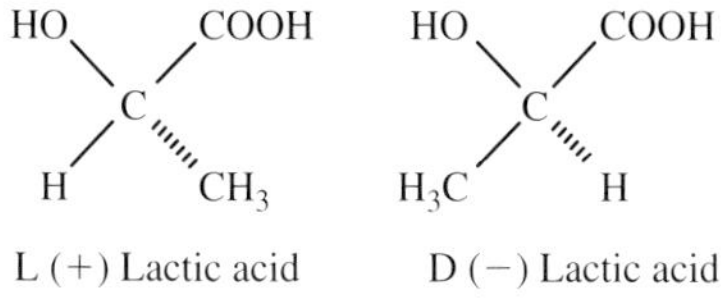

HO COOH HO COOH

L (+) Lactic acid   D (−) Lactic acid

Figure 2-3 Stereoforms of lactic acid.

Lactic acid is the most widely occurring carboxylic acid in nature [7]. It was discovered by the Swedish chemist Scheele in 1780 as a sour component of milk, and was first produced commercially by Charles E. Avery at Littleton, Massachusetts, USA, in 1881. Lactic acid can be manufactured by chemical synthesis or carbohydrate fermentation. First, lactic acid was petrochemically derived [8]. The commercial process for chemical synthesis is based on lactonitrile (CH₃CHOHCN) obtained from acetalaldehyde (CH₃CHO) and hydrogen cyanide (HCN). After recovery and purification by distillation, lactonitrile is then hydrolysed to lactic acid [7, 8]. Lactic acid produced by the petrochemical route exists as a racemic (optically inactive) mixture of D and L forms. Though chemical synthesis produces a racemic mixture, stereo specific lactic acid can be made by carbohydrate fermentation depending on the strain being used.

Lactic acid-based polymers are prepared by polycondensation, ring-opening polymerization and other methods (chain extension, grafting). High molecular weight PLA is generally produced by the ring-opening polymerization of the lactide monomer. The conversion of lactide to high molecular weight polylactide is achieved commercially by two routes. Recently, Cargill Dow used a solvent-free process and a novel distillation process to produce a range of PLA polymers. The process consists of three separate and distinct steps that lead to the production of lactic acid, lactide, and PLA high polymer [8], Fig. 2.4.

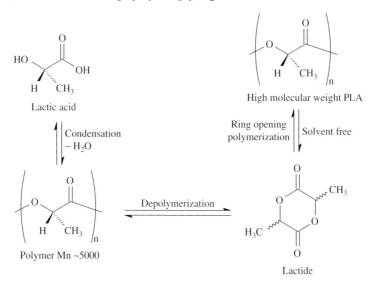

Figure 2-4 Manufacturing route to poly(lactic acid) according to the Cargill Dow process.

Each of the process steps is free of organic solvent: water is used in the fermentation while molten lactide and polymer serve as the reaction media in monomer and polymer production. The essential novelty of the process lies in the ability to go from lactic acid to a low molecular weight polylactic acid, followed by controlled depolymerization to produce the cyclic

dimer, commonly referred to as lactide. This lactide is maintained in liquid form and purified by distillation. Catalytic ring-opening polymerization of the lactide intermediate results in
the production of PLA with controlled molecular weights. The process is continuous with no
necessity to separate the intermediate lactide.

Lactic acid used in the preparation of PLA is derived from annually renewable resources.
Cargill Dow uses sugar from maize as feedstock, due to its low cost and abundance, but it is
envisaged to use local plant sources containing starch, or sugar, such as wheat, sugar beets or
agricultural waste (Fig. 2.5).

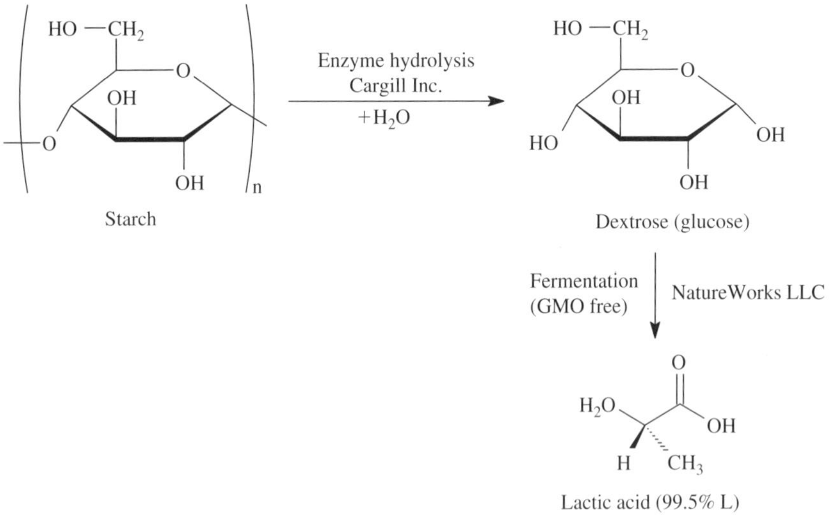

Figure 2-5 Cargill route to lactic acid.

In contrast, Mitsui Toatsu (presently Mitsui Chemicals) utilizes a solvent-based process, in
which a high molecular weight PLA is produced by direct condensation using azeotropic distillation to remove the water of condensation continuously (Fig. 2.6).

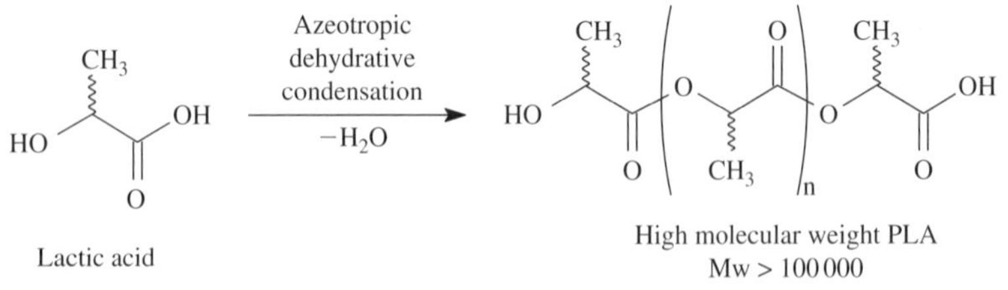

Figure 2-6 Manufacturing route to poly(lactic acid) according to the Mitsui process.

The synthesis of polylactic acid through polycondensation of the lactic acid monomer gave
an average molecular weight lower than 1.6×10^4, whereas ring-opening polymerization
of lactides gave average molecular weights ranging from 2×10^4 to 6.8×10^4 [7]. The ring-
opening polymerization of lactic acid monomers is catalysed by compounds of transition metals:
tin, aluminium, lead, zinc, bismuth, iron and yttrium. Copolymerization and blending of PLA
has been extensively investigated as a useful route to obtain a product with a particular combination of desirable properties. Other ring formed monomers are also incorporated into the lactic

acid-based polymer by ring-opening polymerization [7, 9]. The most utilized comonomers are glycolide (1,4-dioxane-2,5-dione), ε-caprolactone (2-oxepanone), δ-valerolactone (2-pyranone), 1,5-dioxepane-2-one and trimethylene carbonate (1,3-dioxan-2-one). Examples of repeating units of comonomers are given in Table 2.3.

Table 2.3. Repeating units of the most common lactic acid comonomers

Name	Lactones	$\begin{bmatrix} & O \\ & \| \\ O-C-R \end{bmatrix}$ Structure where R
Poly(glycolide)		$-CH_2-$
Poly(lactide)		$\begin{array}{c} CH_3 \\ \| \\ -CH- \end{array}$
Poly (δ-valerolactone)		$-(CH_2)_4-$
Poly (ε-caprolactone)		$-(CH_2)_5-$
Poly(β-hydroxybutyrate)		$\begin{array}{c} CH_3 \\ \| \\ -CH_2-CH- \end{array}$
Poly(β-hydroxyvalerate)		$\begin{array}{c} C_2H_5 \\ \| \\ -CH_2-CH- \end{array}$
Poly(1,5-dioxepane-2-one)		$-(CH_2)_2-O-(CH_2)_2-$
Poly(trimethylene carbonate)		$-O-(CH_2)_3-$

The polymers derived from lactic acid by the polycondensation route are generally referred to as poly(lactic acid) and the ones prepared from lactide by ring-opening polymerization as polylactide [9]. Both types are generally referred to as PLA.

Table 2.4 illustrates PLA polymers commercially available.

Table 2.4. PLA polymers commercially available

Tradename	Supplier	Origin	Website
Lacea	Mitsui Chemicals	Japan	www.mitsui-chem.co.jp/e
Lacty	Shimadzu	Japan	www.shimadzu.co.jp
NatureWorks	Cargill Dow	USA	www.NatureWorksLLC.com
Hycail	Hycail b.v.	The Netherlands	www.hycail.com

2.1.2 Polyhydroxyalkanoates – PHA

Figure 2.7 shows the generic formula for PHAs where x is 1 for all commercially relevant polymers) and R can be hydrogen or hydrocarbon chains of up to C15 in length.

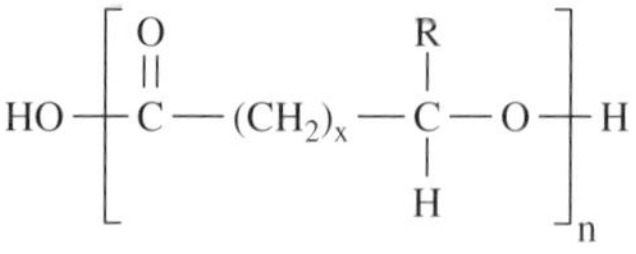

Figure 2-7 Structure of polyhydroxyalkanoates.

Polyhydroxyalkanoates (PHA) are polyesters of various hydroxyalkanoates that are synthesized by many gram-positive and gram-negative bacteria from at least 75 different bacteria [10]. These polymers are accumulated intracellularly to levels as high as 90% of the cell dry weight under conditions of nutrient stress and act as a carbon and energy reserve.

In the 1920s French bacteriologist Lemoigne discovered aliphatic polyester: poly(3-hydroxy butyrate) (PHB) as a granular component in bacterial cells [11]. PHB is the reserve polymer found in many types of bacteria, which can grow in a wide variety of natural environments and which have the ability to produce and polymerize the monomer [R]-3-hydroxybutyric acid. The repeating unit of PHB has a chiral centre (see Fig. 2.8) and the polymer is optically active.

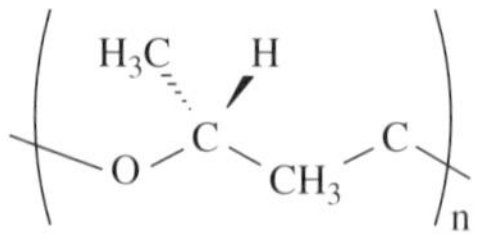

Figure 2-8 Repeating unit of PHB.

It was determined by Stanier, Wilkinson and coworkers that PHB granules in bacteria serve as an intracellular food and energy reserve [11]. PHB polymer is produced by the cell in response to a nutrient limitation in the environment in order to prevent starvation if an essential element becomes unavailable [11]. It is consumed when no external carbon source is available.

Since the discovery of the simple PHB homopolymer by Lemoigne in the mid-1920s, a family of over 100 different aliphatic polyesters of the same general structure has been discovered. PHB is only the parent member of a family of natural polyesters having the same three-carbon backbone structure but differing in the type of alkyl group at the β or 3 position [11]. These polymers are referred to in general as polyhydroxyalkanoates (PHAs) and have the same configuration for the chiral centre at the 3 position, which is very important both for their physical properties and for the activities of the enzymes involved in their biosynthesis and biodegradation. PHAs are also named bacterial polyesters since they are produced inside the cells of bacteria.

A wide range of PHA homopolymers, copolymers, and terpolymers have been produced, in most cases at the laboratory scale. Bacteria that are used for the production of PHAs can be divided into two groups based on the culture conditions required for PHA synthesis [12]. The first group of bacteria requires the limitation of an essential nutrient such as nitrogen, phosphorous, magnesium or sulphur for the synthesis of PHA from an excess carbon source. The following bacteria are included in this group: *Alcaligenes eutrophus*, *Protomonas extorquens* and *Protomonas oleovorans*. The second group of bacteria, which includes *Alcaligenes latus*, a mutant strain of *Azotobacter vinelandii*, and recombinant *Escherichia coli*, do not require nutrient limitation for PHA synthesis and can accumulate polymer during growth.

PHAs exist as discrete inclusions that are typically 0.2 ± 0.5 mm in diameter localized in the cell cytoplasm [12]. The molecular weight of PHAs ranges from 2×10^5 to 3×10^6, depending on the microorganism and the growth conditions.

Today, PHAs are separated into three classes: short chain length PHA (sclPHA, carbon numbers of monomers ranging from C3 to C5), medium chain length PHA (mclPHA, C6–C14), and long chain length PHA (lclPHA, >C14). The main members of the PHA family are the homopolymers poly(3-hydroxybutyrate) (PHB), which has the generic formula in Fig. 2.7 with R = 1(methyl), and poly(3-hydroxyvalerate) (PHV), with the generic formula with R = 2 (ethyl). PHAs containing 3-hydroxy acids have a chiral centre and hence are optically active.

Copolymers of PHAs vary in the type and proportion of monomers, and are typically random in sequence. Poly(3-hydroxybutyrate-*co*-3-hydroxyvalerate) (PHBV) is made up of a random arrangement of the monomers R = 1 and R = 2. Poly(3-hydroxybutyrate-*co*-3-hydroxyhexanoate) (PHBH) consists of the monomers R = 1 (methyl) and R = 3 (propyl). The Nodax® family of copolymers, are poly(3-hydroxybutyrate-*co*-3-hydroxyalkanoate)s with copolymer content varying from 3 to 15% mol% and chain length from C7 up to C19 [13].

Table 2.5. Polyhydroxyalkanoates family

PHA	3-hydroxy acids with side chain R
P(3HB)	—CH_3
P(3HV)	—$CH_2 CH_3$
P(3HB-*co*-3HV) (Biopol®)*	—CH_3 and —CH_2CH_3
P(3HB-*co*-3HHx) (Kaneka)**, (Nodax®)***	—CH_3 and —$CH_2 CH_2 CH_3$
P(3HB-*co*-3HO) (Nodax®)	—CH_3 and —$CH_2 CH_2 CH_2 CH_2CH_3$
P(3HB-*co*-3HOd) (Nodax®)	—CH_3 and —$(CH_2)_{14} CH_3$

* Patent held by Metabolix, Inc.
** Kaneka holds the patent on chemical composition
*** P&G holds processing and application patents

Large-scale commercial production of PHAs uses fermentation technologies. A generic process for PHA produced by bacterial fermentation consists of three basic steps: fermentation, isolation and purification, and blending and palletizing [13]. Subsequent to inoculation and small-scale fermentation, a large fermentation vessel is filled with mineral medium and inoculated with seed ferment (containing the microbe or bacteria). The carbon source is fed at various rates until it is completely consumed and cell growth and PHA accumulation is complete. Current carbon sources for producing PHA: carbohydrates (glucose, fructose, sucrose); alcohols (methanol, glycerol); alkanes (hexane to dodecane); and organic acids (butyrate upwards). In the US, the raw material source is chiefly corn steep liquor; in the EU beet sugar predominates. The total fermentation step typically takes 38 h to 48 h. To isolate and purify PHA, the cells are concentrated, dried and extracted with hot solvent. The residual cell debris is removed from the solvent containing dissolved PHA by a solid–liquid separation process. The PHA is then precipitated by addition of a non-solvent and recovered by the solid–liquid separation process. PHA is washed with solvent to enhance the quality and dried under vacuum and moderate temperatures (in certain cases where high purity product is not needed, solvent extraction may not be required). Separately the solvents are distilled and recycled. The neat polymer is typically preformed into pellets with or without other polymer ingredients [13].

PHAs are produced from a wide variety of substrates such as renewable resources (sucrose, starch, cellulose, triacylglycerols), fossil resources (methane, mineral oil, lignite, hard coal), byproducts (molasses, whey, glycerol), chemicals (propionic acid, 4-hydroxybutyric acid) and carbon dioxide [10].

There are different approaches and pathways for the synthesis of PHAs. Zimm *et al.* [14] distinguished four biosynthethic approaches to produce PHA: *in vitro* via PHA-polymerase catalysed polymerization, and *in vivo* with batch, fed-batch, and continuous (chemostat) cultures.

The biosynthetic pathway of P(3HB) in *A. eutrophus* (now renamed *Ralstonia eutropha*) consists of three enzymatic reactions catalysed by three different enzymes [10, 12] (see Fig. 2.9).

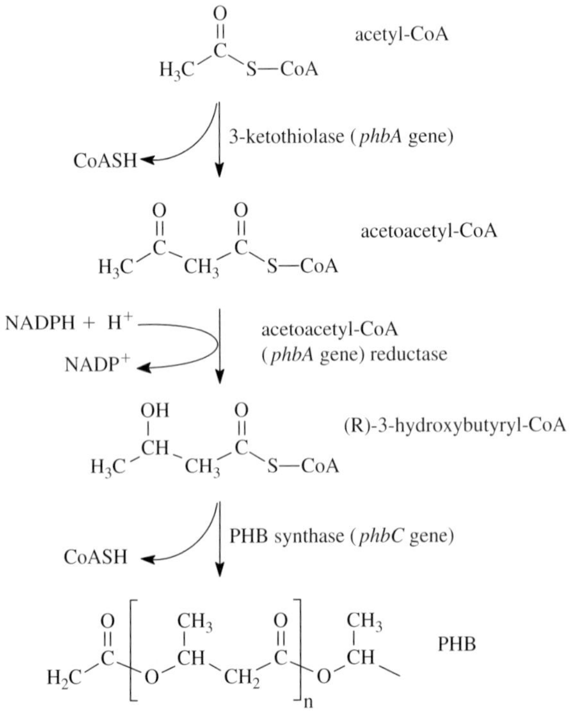

Figure 2-9 PHB synthesis in *Ralstonia eutropha*.

The first reaction consists of the condensation of two acetyl coenzyme A (acetyl-CoA) molecules into acetoacetyl-CoA by β-ketoacylCoA thiolase. The second reaction is the reduction of acetoacyl-CoA to (R)-3-hydroxybutyryl-CoA by an NADPH-dependent acetoacetyl-CoA dehydrogenase. Lastly, the (R)-3-hydroxybutyryl-*co*-A monomers are polymerized into PHB by P(3HB) polymerase.

Homopolymer poly(3-hydroxybutyrate) (PHB) is a brittle, crystalline thermoplastic and undergoes thermal decomposition just at its melting point, thus making processing difficult and limiting its commercial usefulness. Therefore, extensive efforts have been directed towards synthesis of copolymers that have better properties than PHB. Zeneca (formerly Imperial Chemical Industries (ICI)) has developed the PHB copolymer poly(3-hydroxybutyrate-*co*-hydroxyvalerate) (PHBV) also known as Biopol, which is less stiff and less brittle than homopolymer PHB. The ratio of HB to HV monomer can be varied by changing the glucose to propionic acid ratio. By increasing the ratio of HV to HB, the melting temperatures are lower and mechanical properties are improved. In 1996 Zeneca sold its Biopol business to Monsanto, and then in 2001 Metabolix acquired Monsanto Biopol technology. Recently, Metabolix began work on a $15 million programme, supported by the US Department of Energy, to produce PHAs in high yield in native American prairie grass.

Another company, Procter & Gamble, has directed efforts into development and commercialization of a variety of PHA copolymers under the Nodax name. The Nodax® family of copolymers are poly(3-hydroxybutyrate-*co*-3-hydroxyalkanoate)s, with a copolymer content varying from 3 to 15% mol% and chain length from C7 up to C19 [13]. In 2003 Procter & Gamble licensed recovery and processing routes for PHAs to the Japanese company Kaneka Corporation. The companies have a joint agreement to commercialize the Nodax family of PHAs, made from corn or sugar beet and vegetable oils.

Commercially available PHAs are given in Table 2.6.

Table 2.6. PHA polymers commercially available

Tradename	Structure	Supplier	Origin	Website
Biopol®	poly(3-hydroxybutyrate-*co*-3-hydroxyvalerate)	Metabolix	USA	www.metabolix.com
Nodax®	poly(3-hydroxybutyrate-*co*-3-hydroxyalkanoate)s	Kaneka/P&G	Japan	www.nodax.com
Biogreen	poly(3-hydroxybutyrate)	Mitsubishi Gas Chemical	Japan	www.mgc.co.jp
Biomer	poly(3-hydroxybutyrate)	Biomer	Germany	www.biomer.de

2.1.3 Thermoplastic starch – TPS

Starch, the storage polysaccharide of cereals, legumes and tubers, is a renewable and widely available raw material, being the end product of photosynthesis. Starch is composed of a mixture of two substances, an essentially linear polysaccharide-amylose and a highly branched polysaccharide-amylopectin (Fig. 2.10, 2.11).

Both forms of starch are polymers of α-D-glucose. The ratio of both forms varies according to the botanical origin of the starch. Natural starches contain 15–30% amylose and 85–70% amylopectin [15]. Both amylose and amylopectin have a distribution of sizes with different average numbers (degree of polymerization) of glucose residues. The average number of

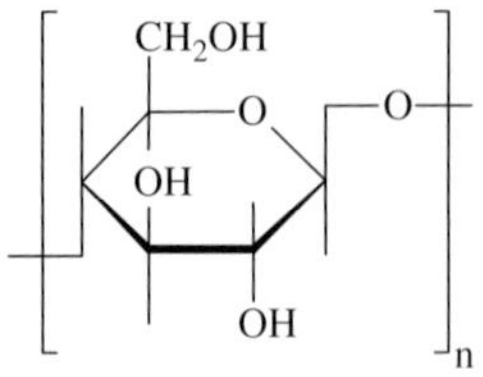

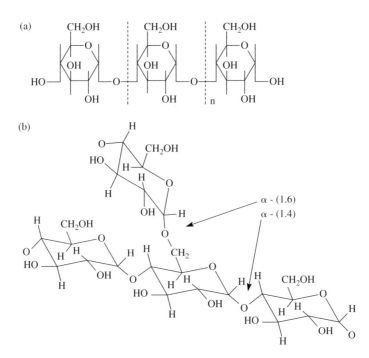

Figure 2-10 General structure of starch.

Figure 2-11 Schematic structure of (a) amylose and (b) amylopectin.

glucose residues for amylose can vary from 250 to 5000, and the average number of glucose residues for amylopectin can vary from 10 000 to 100 000.

Amylose is a relatively long, linear α-glucan containing around 99% $(1 \rightarrow 4)$-α- and 1% $(1 \rightarrow 6)$-α-linkages [16]. Amylose has a molecular weight of approximately 1×10^5–1×10^6, a degree of polymerization (DP) by number (DP_n) of 324–4920 with around 9–20 branch points equivalent to 3–11 chains per molecule. Amylopectin is a much larger than amylose with a molecular weight of 1×10^7–1×10^9 and a heavily branched structure built from about 95% $(1 \rightarrow 4)$-α- and 5% $(1 \rightarrow 6)$-α-linkages. The DP_n is typically within the range 9600–15 900.

The size, shape, and morphology of the starch granules are characteristic of the particular botanical source (Fig. 2.12). Starch granules, typically ranging in size from 2 to 30 μm, depending on the plant origin, are partially crystalline and insoluble in cold water.

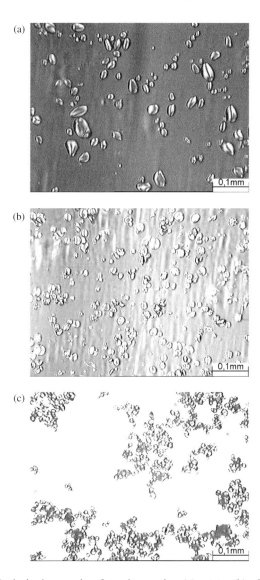

Figure 2-12 Optical micrographs of starch granules: (a) potato, (b) wheat, (c) maize.

Table 2.7. Diameter and gelatinization temperature of starch granules [17]

Source	Mean diameter, μm	Gelatinization temperature, °C
Corn	15	62–71
Wheat	20–22	53–64
Rice	5	65–73
White potato	33	62–68
Sweet potato	25–50	82–83
Tapioca	20	59–70

The conventional processing of starch, including food processing and processing to produce pastes, thickeners and adhesives, is in the presence of heat and excess water [18].

In 1980s a breakthrough occurred by processing starch at approximately its natural water content (15%) in a closed volume at temperatures above 100°C. Using conventional injection moulding, glassy, amorphous, thermoplastic starch (TPS) polymers (Tg 60°C) were obtained with moduli similar to those of polypropylene and high-density polyethylene.

Thermoplastic starch can be produced from native starch using a swelling or plasticizing agent while applying a dry starch in compound extruders without adding water. When a starch with a water content higher than 5% is plastified or pasted under pressure and temperature, a de-structured starch is always formed. In the production procedure of thermoplastic starch, the mainly water-free raw material is homogenized and melted in an extrusion process with a plastifing material. Several plasticizers have been studied, including water, glycerol, sorbitol, glycol, poly(ethylene glycol), urea, glucose, maltose, as well as melt-flow accelerators, such as lecitin, glycerol monostearate, and calcium stearate [19].

The glass temperature of starch-containing materials is a function of plasticizer content. Depending on the processing conditions and plasticizer content, thermomechanical processing of granular starch with the aid of plasticizers and melt-flow accelerators gives a complex starch, plastic material. This is composed of residual swollen granular starch, partially melted, deformed and disrupted granules, completely molten starch, and recrystallized starch. The degree of disruption and melting of the various granular starches is regulated by the plasticizer content and by the processing parameters (shear stress, melt viscosity and temperature).

The starch destructurization is defined as a partial fragmentation of the crystalline structure within the polysaccharides. By the transformation of native starch materials to highly amorphous thermoplastics, the compounded thermoplastic starch (TPS) formulation is re-meltable and extrusion or injection moulding is processable by renewing the energy input. Native starches can be destructurized within co-rotating twin screw extruder systems by a controlled feeding of suitable destructurization additives (water, glycerol) in combination with defined operating parameters [20].

Blends or composites materials have been produced by the processing of starch with biodegradable polymers such as poly(ε-caprolactone), poly(lactic acid), poly(vinyl alcohol), poly(hydroxybutyrate-*co*-valerate), and polyesteramide. The most common are Mater-Bi from Novamont and Ecostar from National Starch.

Table 2.8. Starch-based polymers commercially available

Tradename	Structure	Supplier	Origin	Website
Solanyl	Starch based	Rodenburg Biopolymers	The Netherlands	www.biopolymers.nl
Bioplast TPS	Thermoplastic starch	Biotem	Germany	www.biotec.de
EverCorn	Starch based	Japan Corn Starch	Japan	www.japan-cornstarch.com
Plantic	Starch based	Plantic Technologies	Australia	www.plantic.com.au
Biopar	Starch based	BIOP Biopolymer Technologies AG	Germany	www.biopag.de
Placorn	Starch based	Nihon Shokuhin Kako	Japan	www.nisshoku.co.jp

2.2. OTHER COMPOSTABLE POLYMERS FROM RENEWABLE RESOURCES

2.2.1 Cellulose

Cellulose, the most abundant organic compound on earth, is the major structural component of the cell wall of higher plants [15]. It is major component of cotton (95%), flax (80%), jute (60–70%) and wood (40–50%). Cellulose pulps can be obtained from many agricultural by-products such as sugarcane, sorghum bagasse, corn stalks, and straws of rye, wheat, oats, and rice.

Cellulose is a polydisperse linear polysaccharide consisting of β-1,4-glycosidic linked D-glucose units (so-called anhydroglucose unit) (see Fig. 2.13).

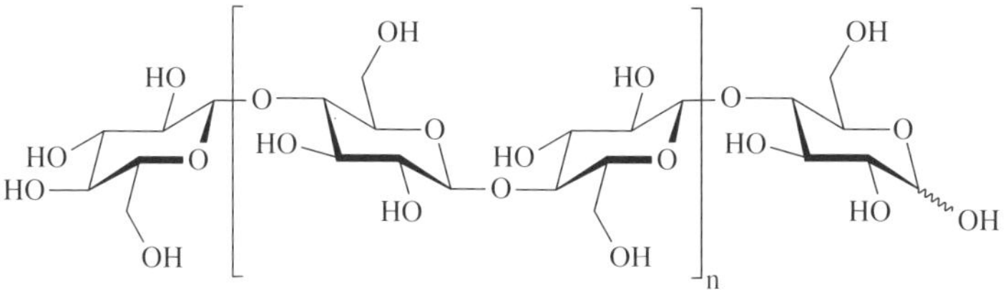

cellulose chain, where n = 2000 to 10 000

Figure 2-13 Schematic structure of cellulose.

The consequence of the supra-molecular structure of cellulose is its insolubility in water, as well as in common organic liquids [15, 21]. Poor solubility in common solvents is one of the reasons why cellulose is converted to its cellulose esters. Another reason is that cellulose is not melt-processible, because it decomposes before it undergoes melt flow [22].

Cellulose esters have been commercially important polymers for nearly a century, and have found a variety of applications, including solvent-borne coatings, separation, medical and controlled release applications as well as composites and laminates and plastics.

The most common cellulose esters comprise cellulose acetate (CA), cellulose acetate propionate (CAP), and cellulose acetate butyrate (CAB). They are thermoplastic materials produced through esterification of cellulose. Different raw materials such as cotton, recycled paper, wood cellulose, and sugarcane are used to make the cellulose ester biopolymers in powder form. Bioceta, plasticized cellulose acetate, is prepared from cotton flakes and wood pulp through an esterification process with acetic anhydride. Cellulose acetate propionate (CAP) and cellulose acetate butyrate (CAB) are mixed esters produced by treating cellulose with appropriate acids and anhydrides in the presence of sulphuric acid.

Cellulose-based polymers are given in Table 2.9.

Table 2.9. Cellulose-based polymers commercially available

Tradename	Structure	Supplier	Origin	Website
Natureflex	Cellulose based	Innovia Films (formerly Surface Specialties-UCB)	UK	www.innoviafilms.com
Tenite	Cellulose esters	Eastman	USA	www.eastman.com
Bioceta	Cellulose acetate	Mazzucchelli	Italy	www.mazzucchelli1849.it
Cellidor	Cellulose acetate propionate; cellulose acetate butyrate	Albis Plastics	Germany	www.albis.com

2.2.2 *Chitosan*

Chitin (poly(N-acetyl-D-glucosamine)) represents the second most abundant polysaccharide after cellulose. It is found in the exoskeleton of crustaceans and insects and in the cell wall of fungi and microorganisms [23]. Arthropod shells (exoskeletons), the most easily accessible sources of chitin, contain 20–50% of chitin on a dry basis. Wastes of seafood processing industries are used for the commercial production of chitin.

The structure of chitin is essentially the structure of cellulose, with the hydroxyl group at C-2 of the D-glucopyranose residue substituted with an N-acetylamino group [15] (see Fig. 2.14).

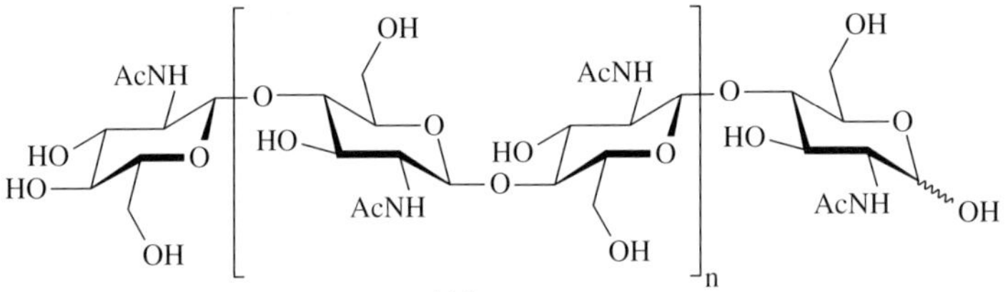

Figure 2-14 Schematic structure of chitin.

Chitosan, poly-β(1,4)-2-amino-2-deoxy-D-glucopyranose, is the deacetylated product of chitin (Fig. 2.15).

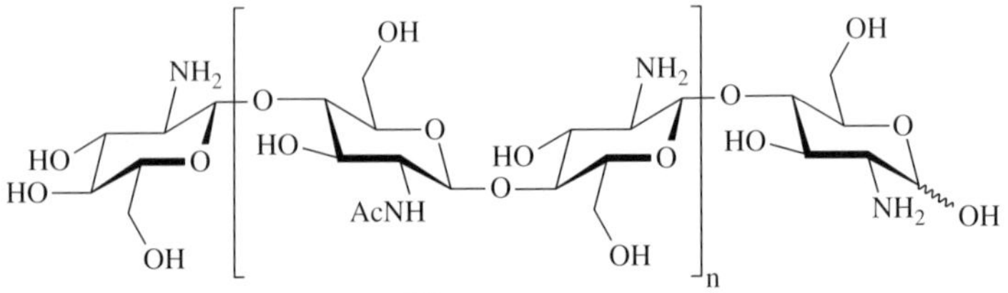

where m is <<< n

Figure 2-15 Schematic structure of chitosan.

Chitosan is composed of glucosamine (2-amino-2-deoxy-glucopyranose) and N-acetyl glucosamine (2-acetamido-2-deoxy-glucopyranose) linked in a β (1,4)-manner; the glucosamine to N-acetyl glucosamine ratio being referred to as the degree of deacetylation [24]. Depending on the source and preparation procedure, its molecular weight may range from 300 to over 1000 kD with degrees of deacetylation from 30% to 95%.

Chitosan is obtained on an industrial scale by the alkaline deacetylation of chitin [23, 24]. The main commercial sources of chitin are shells of shellfish (mainly crabs, shrimps, lobsters and krills), wastes of the seafood processing industry. Basically, the process consists of deproteinization with a dilute NaOH solution, demineralization with a dilute HCl solution and decolouration of the raw shell material. Chitin is obtained as an almost colourless to off-white powdery material. Chitosan is produced by deacetylating chitin using 40–50% aqueous alkali at 100–160°C for a few hours. The resultant chitosan has a degree of deacetylation up to 0.95.

There are many producers of chitin and chitosan in the world; Table 2.10 gives producers found in Europe.

Table 2.10. Chitosan producers in Europe

Structure	Supplier	Origin	Website
Chitosan	France Chitine	France	www.france-chitine.com
Chitosan	Nova Matrix	Norway	www.novamatrix.biz
Chitosan	Primex	Iceland	www.primex.is
Chitosan	Heppe GmbH	Germany	www.biolog-heppe.de

2.2.3 Proteins

A protein is considered to be a random copolymer of amino acids. A generic protein monomeric unit is given in Fig. 2.16, where R represents the side chain of an amino acid. Proteins
can be divided into proteins from plant origin (e.g. gluten, soy, pea and potato) and proteins
from animal origin (e.g. collagen (gelatin), casein, silk, keratin, whey).

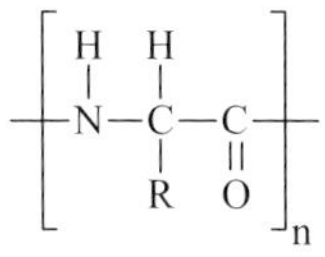

Figure 2-16 Schematic structure of proteins.

Potential candidates for use in the fabrication of biodegradable films include soy proteins,
wheat gluten, corn proteins, myofibrillar proteins from fish and pea proteins [25, 26]. Proteins
are considered as structured heteropolymers [26]. Two classes of proteins can be distinguished,
globular or pseudoglobular proteins such as globulins or gliadins and fibrous or "polymerized"
proteins such as collagen or glutenins.

Gluten is a mixture of monomeric proteins (gliadins) and polymerized proteins (glutenins) linked through intermolecular disulphide bridges. Gluten is the main storage protein in
wheat. In general, gluten-based plastics require the addition of plasticizer agents. Hydrophilic
compounds (water, polyols, oligosaccharides), lipidic compounds (waxes, oils, fatty acids,
monoglycerols) are used as protein plasticizers, the most frequently used is glycerol [27, 28].
Plasticizers decrease the protein interactions and increase polymer chain mobility and intermolecular spacing, decreasing also the glass transition temperature of proteins.

Soy protein-based plastics are another group of biodegradable, environmentally friendly,
polymer materials from an abundantly renewable resource [29–31]. There are several types of
soybean products that can potentially be utilized for engineering structural applications [29].

Two processes are currently used to prepare protein-based films: the wet method ("casting"), which involves the solubilization of protein and a plasticizer in a solvent followed by
the formation of a protein network on evaporation of the solvent; and the dry method, which is
based on thermoplastic characteristics of proteins and combines the use of pressure and heat to
plasticize protein chains [25, 32]. Dehulled soybean, after solvent defatting and meal grinding,
becomes a fat-free, low fibre soy flour (48.5% protein). The soy flour, after leaching out of
the water/alcohol soluble sugars, is termed soy protein concentrate (above 65% protein). The
soy protein concentrate, if it is further extracted by alkali and reprecipitated by acidification,
becomes the purest commercially available soy protein isolate (above 90% protein).

Vegetable and animal proteins have been used in many non-food applications, but despite
the potential, protein-based plastics have not yet made significant progress in commercialization at a large scale.

2.3. BIODEGRADABLE POLYMERS FROM PETROCHEMICAL SOURCES

Aliphatic polyesters are the representatives of synthetic biodegradable polymers.

Synthetic biodegradable polyesters are generally made by the polycondensation method and raw materials are obtained from petrochemical feed stocks. Aliphatic polyesters such as poly(butylene succinate) and poly(ε-caprolactone) are commercially produced. Besides these aliphatic polyesters, various types of synthethic biodegradable polymers have been designed [33]. They are, for example, poly(ester amide)s, poly(ester carbonate)s, poly(ester urethane)s, etc. Most of them are still at a premature stage.

The traditional way of synthesizing polyesters has been by polycondensation using diols and a diacid (or an acid derivative), or from a hydroxy acid [33, 34].

Polycondensation can be applicable for a variety of combinations of diols and diacids, but it requires, in general, higher temperature and longer reaction time to obtain high molecular weight polymers. In addition, this method suffers from such shortcomings as the need for removal of reaction by-products and a precise stoichiometric balance between reactive acid and hydroxy groups. The ring-opening polymerization (ROP) of lactones, cyclic diesters (lactides and glycolides), is an alternative method, which can be carried out under milder conditions to produce high molecular weight polymers in a shorter time. Furthermore, recent progress in catalysts has enabled the production of polyesters of controlled chain lengths.

Recently, enzyme-catalysed polymer synthesis has been established as another approach to biodegradable polymer preparation [35–37].

2.3.1 Aliphatic polyesters and copolyesters

One of the most promising polymers in this family is poly(butylene succinate) (PBS), which is chemically synthesized by the polycondensation of 1,4-butanediol with succinic

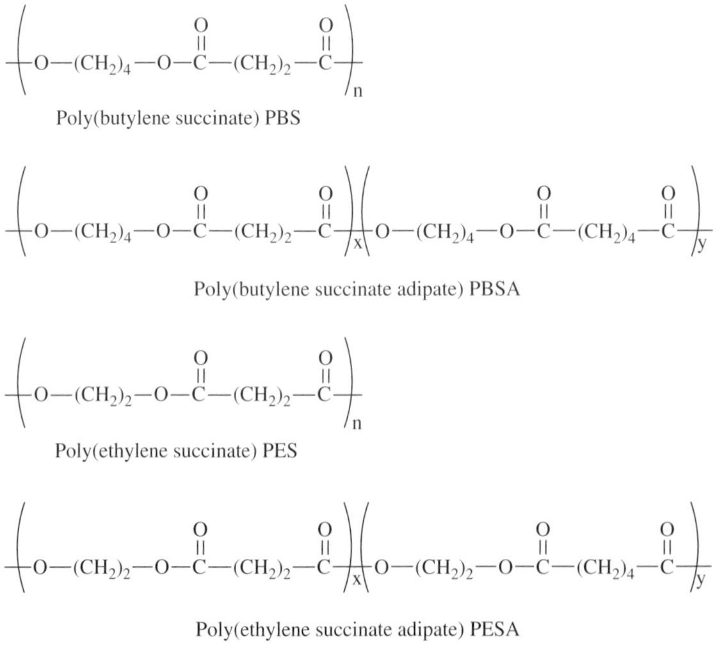

Poly(butylene succinate) PBS

Poly(butylene succinate adipate) PBSA

Poly(ethylene succinate) PES

Poly(ethylene succinate adipate) PESA

Figure 2-17 Aliphatic polyesters and copolyesters.

acid. High molecular weight PBS is generally prepared by a coupling reaction of relatively low molecular weight PBS in the presence of hexamethylene diisocyanate as a chain extender.

Bionolle is produced through the polycondensation reaction of glycols such as ethylene glycol and butanediol-1,4, and aliphatic dicarboxylic acids such as succinic and adipic acid used as principal raw materials [38]. Aliphatic polyesters, trademarked "Bionolle", such as polybutylene succinate (1000 series), polybutylene succinate adipate copolymer (3000 series) and polyethylene succinates (6000 series), with high molecular weights ranging from several tens of thousands to several hundreds of thousands, were invented in 1990 and produced through the polycondensation reaction of glycols with aliphatic dicarboxylic acids and others.

Commercially available aliphatic polyesters and copolyesters are given in Table 2.11.

Table 2.11. Aliphatic polyesters and copolyesters commercially available

Tradename	Supplier	Origin	Website
Bionolle® 1000 Poly(butylene succinate) PBS	Showa Highpolymer	Japan	www.shp.co.jp
Bionolle® 2000 Bionolle® 3000 Poly(butylene succinate adipate) PBSA	Showa Highpolymer	Japan	www.shp.co.jp
Bionolle 6000® Poly(ethylene succinate) PES	Showa Highpolymer	Japan	www.shp.co.jp
Bionolle 7000® Poly(ethylene succinate adipate) PESA	Showa Highpolymer	Japan	www.shp.co.jp
SkyGreen SG 100 Poly(butylene succinate) PBS SG200 Poly(butylene succinate adipate) PBSA	SK Polymers	Korea	www.skchemicals.com/english

2.3.2 Aromatic polyesters and copolyesters

While the biological susceptibility of many aliphatic polyesters has been known for many years, aromatic polyesters such as polyethylene terephthalate (PET) or polybutylene terephthalate are regarded as non-biodegradable [39]. To improve the use properties of aliphatic polyesters, an attempt was made to combine the biodegradability of aliphatic polyesters with the good material performance of aromatic polyesters in novel aliphatic–aromatic copolyesters.

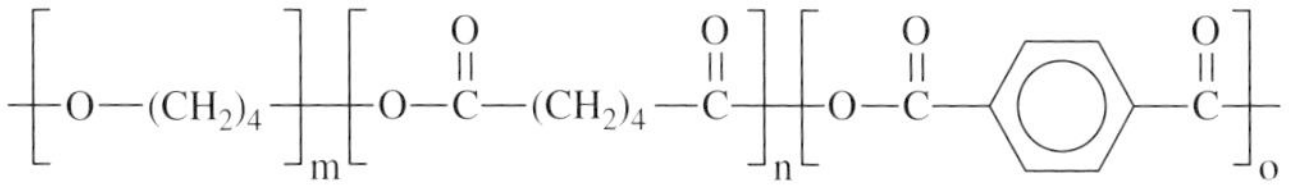

Figure 2-18 Aromatic copolyesters.

Using standard polycondensation techniques, copolyesters with molar masses in a range necessary for technical application were obtained [40–41]. The best results with regard to the use properties were achieved with a combination of 1,4-butanediol, adipic acid and terephthalic acids.

Commercially available aromatic copolyesters are given in Table 2.12.

Table 2.12. Aromatic copolyesters commercially available

Tradename	Supplier	Origin	Website
Biomax® Poly(butylene succinate terephthalate) PBST	DuPont	USA	www.dupont.com
Eastar Bio® Poly(butylene adipate terephthalate) PBAT	Eastman Chemicals*	Japan	www.eastman.com
Ecoflex® Poly(butylene adipate terephthalate) PBAT	BASF	Germany	www.basf.com

* In 2004 Eastar Bio technology was bought by Novamont.

Poly(trimethylene terephthalate) (PTT) is a linear aromatic polyester produced by polycondensation of 1,3-propanediol (trimethylene glycol or PDO) with either purified terephthalic acid (PTA) or trimethyl terephthalate (DMT) (Fig. 2.19).

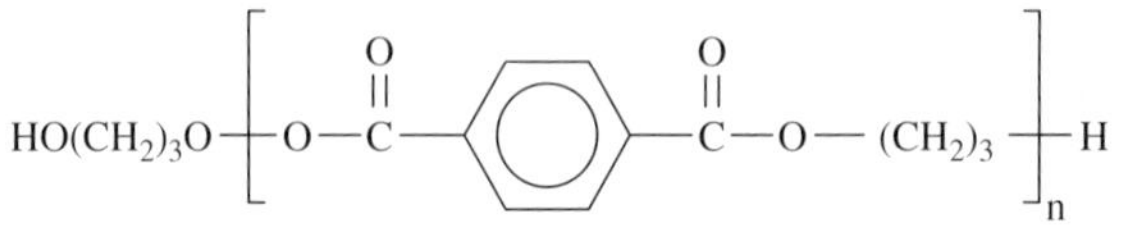

Figure 2-19 Schematic structure of poly(trimethylene terephthalate).

While both these monomers – the diacid and the diol component – are conventionally derived from petrochemical feedstocks, DuPont, Tate & Lyle and Genecor have recently succeeded in introducing PDO using an aerobic bioprocess with glucose from corn starch as the feedstock, opening the way for bulk production of PTT from a bio-based monomer.

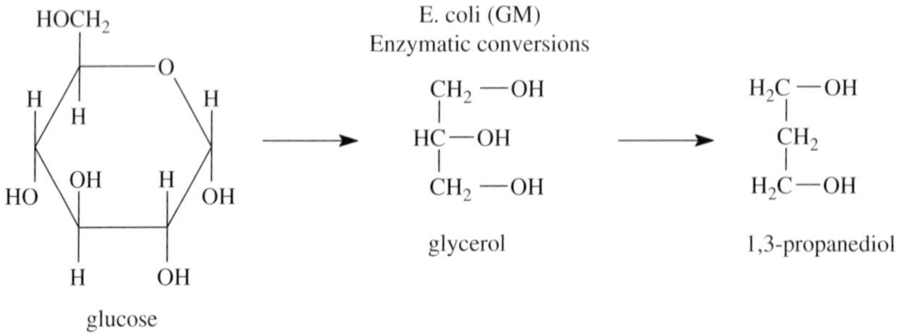

Figure 2-20 Biotechnological route to 1,3-propanediol.

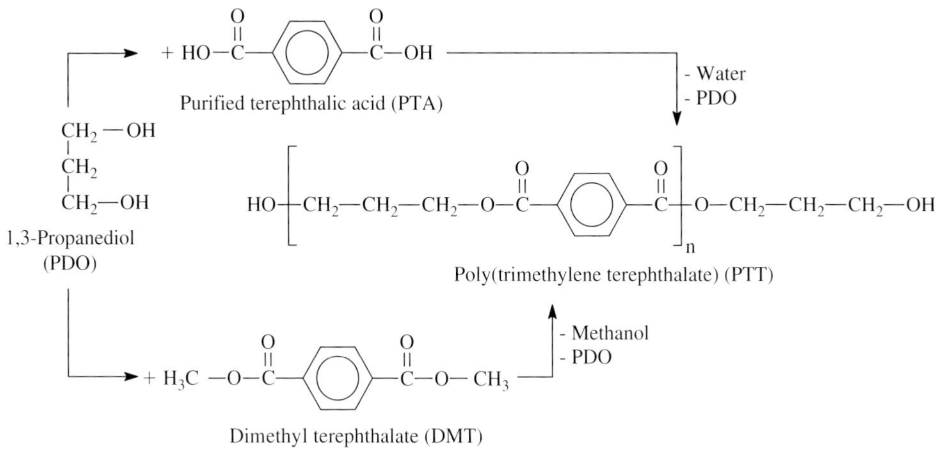

Figure 2-21 Manufacturing routes to poly(trimethylene terephthalate).

The natural fermentation pathway to PDO involves two steps: yeast first ferments glucose to glycerol, then bacteria ferment this to PDO. In the bioprocess developed by DuPont, dextrose derived from wet-milled corn is metabolized by genetically engineered *E. coli* bacteria and converted within the organism directly to PDO via an aerobic respiration pathway (Fig. 2.20). The PDO is then separated from the fermentation broth by filtration, and concentrated by evaporation, followed by purification by distillation. The PDO is then fed to the polymerization plant. PTT can be produced by transesterification of dimethyl terephthalate (DMT) with PDO, or by the esterification route, starting with purified terephthalic acid (PTA) and PDO (Fig. 2.21) [13]. The polymerization can be a continuous process and is similar to the production of PET. In the first stage of polymerization, low molecular weight polyester is produced in the presence of excess PDO, with water of esterification (in the case of PTA) or methanol (in the case of DMT) being removed. In the second stage, polycondensation, chain growth occurs by removal of PDO and remaining water/methanol. As chain termination can occur at any time (due to the presence of a monofunctional acid or hydroxyl compound), both monomers must be very pure. As the reaction proceeds, removal of traces of PDO becomes increasingly difficult. This is compensated for by having a series of reactors operating under progressively higher temperatures and lower pressures. In a final step, highly viscous molten polymer is blended with additives in a static mixer and then pelletized.

Table 2.13 summarizes commercially available poly(trimethylene terephthalate) polymers.

Table 2.13. Poly(trimethylene terephthalate) (PTT) polymers

Tradename	Supplier	Origin	Website
Sorona™	DuPont	USA	www.dupont.com
Corterra®	Shell	Canada	www.shellchemicals.com
PermaStat	RTP	USA	www.rtpcompany.com

2.3.3 Poly(caprolactone) – PCL

Poly(ε-caprolactone) (PCL) is a linear polyester manufactured by ring-opening polymerization of a seven-membered lactone, ε-caprolactone.

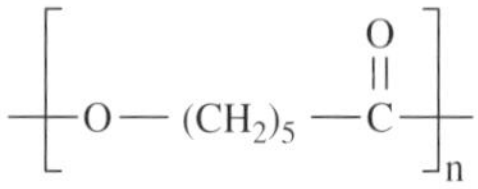

Figure 2-22 Structure of poly(ε-caprolactone).

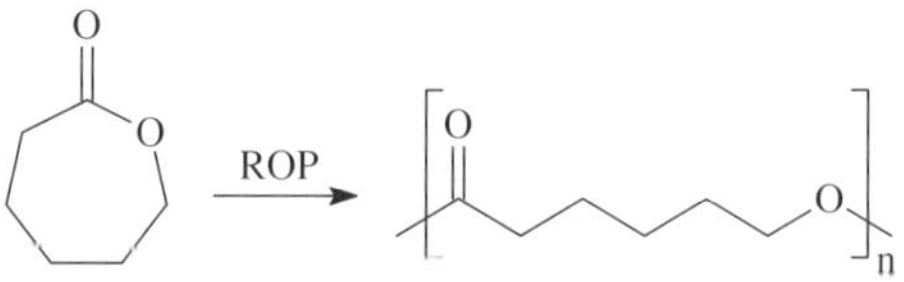

Figure 2-23 Schematic route to PCL.

Anionic, cationic, coordination, or radical polymerization routes are all applicable [42, 42]. Recently, enzymatic catalysed polymerization of ε-caprolactone has been reported [36]. It is a semicrystalline polymer with a degree of crystallinity around 50%. It has a rather low glass transition temperature ($-60°C$) and melting point ($61°C$).

PCL was recognized as a biodegradable and non-toxic material, and a promising candidate for controlled release applications, especially for long-term drug delivery. It may be copolymerized with many other lactones, such as glycolide, lactide, δ-valerolactone, ε-decalactone, poly(ethylene oxide), and alkyl-substituted ε-caprolactone. Blends of PCL with other biodegradable polymers such as PHB, PLA, and starch have been prepared.

Table 2.14. PCL polymers commercially available

Tradename	Supplier	Origin	Website
Tone	Union Carbide	USA	www.unioncarbide.com
CAPA	Solvay	Belgium	www.solvay.com
Placcel	Daicel Chemical Indus.	Japan	www.daicel.co.jp/english/kinouhin/category/capro.html

2.3.4 Poly(esteramide) – PEA

Polyesteramide BAK 1095 is based on caprolactam (Nylon 6), butanediol and adipic acid; BAK 2195 is based on adipic acid and hexamethylene-diamine (Nylon 6,6) and adipic acid with butanediol and diethylene glycol as ester components [44]. The production process is solvent and halogen free.

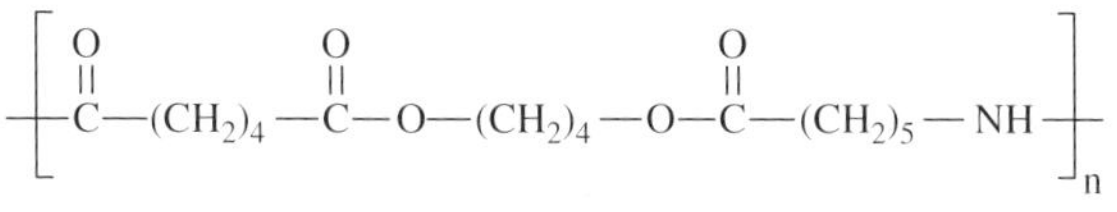

Figure 2-24 Schematic structure of poly(esteramide)s.

Table 2.15. Polyesteramides commercially available

Tradename	Supplier	Origin	Website
BAK*	Bayer AG	Germany	www.bayer.com

* In 2002 the production was suspended.

2.3.5 Poly(vinyl alcohol) – PVA

Poly (vinyl alcohol) (PVA) (Fig. 2.25) is the largest volume water-soluble polymer produced today. PVA is not produced by direct polymerization of the corresponding monomer, since vinyl alcohol tends to convert spontaneously into the enol form of acetalaldehyde, driven by thermodynamic reasons and with extremely limited kinetic control [44]. PVA is attained instead from the parent homopolymer poly(vinyl acetate) (PVAc). The polymerization of vinyl acetate occurs via a free-radical mechanism, usually in an alcoholic solution (methanol, ethanol) although for some specific applications a suspension polymerization can be used. The scheme for industrial production of PVA is given in Fig. 2.26.

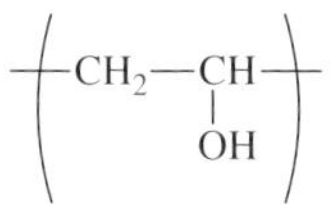

Figure 2-25 Schematic structure of PVA.

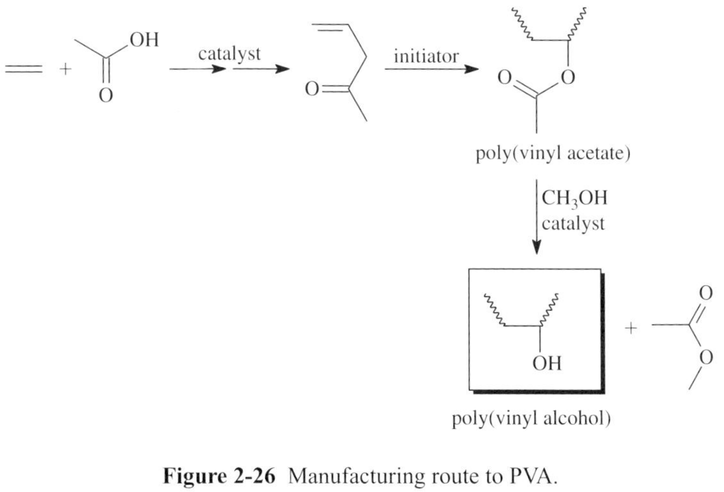

Figure 2-26 Manufacturing route to PVA.

PVA is produced on an industrial scale by hydrolysis (methanolysis) of PVAc, often in a one-pot reactor. Different grades of PVA are obtained depending upon the degree of hydrolysis (HD). Polymerization reactions can be carried out in batch or in continuous processes, the latter being used mostly for large-scale production. In the continuous industrial process, the free-radical polymerization of vinyl acetate is followed by alkaline alcoholysis of PVAc. The molecular weight of PVAc is usually controlled by establishing the appropriate residence time in the polymerization reactor, vinyl acetate feed rate, solvent (methanol) amount, radical initiator concentration, and polymerization temperature.

The main producers of PVA are given in Table 2.16.

Table 2.16. Poly(vinyl alcohol)s producers

Tradename	Supplier	Origin	Website
Mowiol	Clariant GmbH	Germany	www.cepd.clarinet.com
Erkol	Erkol SA	Spain	www.erkol.com
Sloviol	Novacky	Slovakia	www.nchz.sk
Polyvinol	Vinavil SpA	Italy	www.mpaei.it/it/vinavil/home.htm
Elvanol	DuPont	USA	www.dupont.com/industrial-polymers/ elvanol/index.html
Cevol	Celanep	USA	www.celanesechemicals.com
Airvol	Air Products	USA	
Kuraray Poval	Kuraray Co. Ltd	Japan	www.kuraray.co.jp/en
Unitika Poval	Unitika Ltd	Japan	www.unitika.co.jp/e/home_e2.htm
Gohsenol	Nippon Gohsei – The Nippon Synthetic Chemical Industry Co. Ltd	Japan	www.nippongohsei.com/gohsenol/ index.htm
Hapol	Hap Heng	China	

2.3.6 Blends

One of the strategies adopted in producing compostable polymer materials is blending of biodegradable polymers. Blending is a common practice in polymer science to improve unsatisfactory physical properties of the existing polymer or to decrease cost. By varying the composition and processing of blends, it is possible to manipulate properties. The leading compostable blends are starch-based materials. The aim is to combine the low cost of starch with higher cost polymers having better physical properties. An example of such material is Mater-Bi manufactured by Novamont [46]. Mater-Bi is prepared by blending starch with other

Table 2.17. Commercially available blends

Tradename	Supplier	Origin	Website
Mater-Bi	Novamont	Italy	www.materbi.com
Ecostar	National Starch	USA	www.nationalstarch.com
Ecofoam	National Starch	USA	www.nationalstarch.com
Biograde (cellulose blends)	FKuR	Germany	www.fkur.de
Bioflex (PLA blends)	FKuR	Germany	www.fkur.de
Fasal (celluse based)	Austel + IFA	Austria	www.austel.at
Cereplast	Cereplast, Inc.	USA	www.cereplast.com

biodegradable polymers in an extruder in the presence of water or plasticizer. Three main classes of Mater-Bi are commercially available:

- Class Z – thermoplastic starch and polycaprolactone.
- Class Y – thermoplastic starch and cellulose derivatives.
- Class V – thermoplastic starch more than 85%.

REFERENCES

[1] Narayan R., Doi Y., Fukada K.: Impact of Government Policies, Regulations, and Standards Activities on an Emerging Biodegradable Plastics Industry. Biodegradable Plastics and Polymers. Elsevier, New York, 1994, p. 261.

[2] ASTM D 6400-04 "Standard Specification for Compostable Plastics".

[3] ISO/DIS 17088 – Specifications for compostable plastics.

[4] EN ISO 472:2001 – Plastics-Vocabulary.

[5] www.bpsweb.net/02_english

[6] Stevens E.S.: How green are green plastics? *BioCycle* 43 (2002) 42.

[7] Narayanan N., Roychoudhury P.K., Srivastava A.: L(+) lactic acid fermentation and its product polymerization, *Electronic Journal of Biotechnology* 7 (2004) 167.

[8] Lunt J.: Large-scale production, properties and commercial applications of polylactic acid polymers, *Polym. Deg. Stab.* 59 (1998) 145.

[9] Södergård A., Stolt M.: Properties of lactic acid based polymers and their correlation with composition, *Prog. Polym. Sci.* 27 (2002) 1123.

[10] Reddy C.S.K., Ghai R., Rashmi, Kalia V.C.: Polyhydroxyalkanoates: an overview, *Bioresource Technology*, 87 (2003) 137.

[11] Lenz R.W., Marchessault R.H.: Bacterial polyesters: biosynthesis, biodegradable plastics and bio-technology, *Biomacromolecules* 6 (2005) 1.

[12] Khanna S., Srivastava A.K.: Recent advances in microbial polyhydroxyalkanoates, *Process Biochemistry* 40 (2005) 607.

[13] Techno-economic feasibility of large-scale production of bio-based polymers in Europe (PRO-BIP), Final Report, Utrecht/Karlsruhe, October 2004.

[14] Zinn M., Witholt B., Egli T.: Occurrence, synthesis and medical application of bacterial polyhy-droxyalkanoates, *Adv. Drug Deliv. Rev.* 53 (2001) 5.

[15] Robyt J.F.: *Essentials of Carbohydrate Chemistry*, Springer-Verlag, New York, 1998.

[16] Tester R.F., Karkalas J., Qi X.: Starch-composition, fine structure and architecture, *J. Cer. Sci.* 39 (2004) 151.

[17] Daniel J.R., Voragen A.C.J., Pilnik W.: Starch and other polysaccharides in: *Ullmann's Encyclopedia of Industrial Chemistry*, Vol. A25, 1994 VCH Verlagsgesellschaft.

[18] Ross-Murphy S.B., Stepto R.T.: Greening polymers for the 21st century: real prospects and virtual realities, in: *Emerging Themes in Polymers Science*, Royal Society of Chemistry, London, 2001.

[19] Van Soest J.J.G., Vliegenthart J.F.G.: Crystallinity in starch plastics: consequences for material properties. TIBTECH 15 (1997) 208.

[20] Aichholzer W., Fritz H.-G.: Rheological characterization of thermoplastic starch materials, *Starch* 50 (1998) 77.

[21] Heinze T., Liebert T.: Unconventional methods in cellulose functionalisation, *Prog. Polym. Sci.* 26 (2001) 1605.

[22] Edgar K.J., Buchanan C.M., Debenham J.S., Rundquist P.A., Seiler B.D., Shelton M.C., Tindall D.: Advances in cellulose ester performance and application, *Prog. Polym. Sci.* 26 (2001) 1605.

[23] Kurita K.: Controlled functionalisation of the polysaccharide chitin, *Prog. Polym. Sci.* 26 (2001) 1921.

[24] Di Martino A., Sittinger M., Risbud M.V.: Chitosan: a versatile biopolymer for orthopaedic tissue engineering, *Biomaterials* 26 (2005) 5983.

[25] Orliac O., Rouilly A., Silvestre F., Rigal L.: Effects of additives on the mechanical properties, hydrofobicity and water uptake of thermo-moulded films produced from sunflower isolate, *Polymer* 43 (2002) 5417.

[26] Larré C., Desserme C., Barbot J., Guéguen J.: Properties of deamidated gluten films enzymatically cross-linked, *J. Agri. Food Chem.* 48 (2000) 5444.

[27] Pommet M., Redl A., Morel M.-H., Guilbert S.: Study of wheat gluten plasticization with fatty acids, *Polymer* 44 (2003) 115.

[28] Sánchez A.Ch., Popineau Y., Mangavel C., Larré C., Guéguen J.: Effect of different plasticizers on the mechanical and surface properties of wheat gliadin films, *J. Agric. Food Chem.* 46 (1998) 4539.

[29] Sue H.-J., Weng S., Jane J.-L.: Morphology and mechanical behaviour of engineering soy plastics, *Polymer* 38 (1997) 5036.

[30] Zhang J., Mungara P., Jane J.: Mechanical and thermal properties of extruded soy protein sheets, *Polymer* 42 (2001) 2569.

[31] Lodha P., Netravali A.N.: Thermal and mechanical properties of environment-friendly "green" plastics from stearic acid modified-soy protein isolate, *Industrial Crops and Products* 21 (2005) 49–64.

[32] Micard V., Morel M.-H., Bonicel J., Guibert S.: Thermal properties of raw and processed wheat gluten in relation with protein aggregation, *Polymer* 42 (2001) 477.

[33] Okada M.: Chemical syntheses of biodegradable polymers, *Prog. Polym. Sci.* 27 (2002) 87.

[34] Albertsson A.-Ch., Varma I.K.: Aliphatic polyesters: synthesis, properties and applications in: Degradable aliphatic polyesters, *Adv. Polym. Sci.* Vol. 157, Springer (2002).

[35] Namekawa S., Suda S., Uyama H., Kobayashi S.: Lipase-catalysed ring-opening polymerization of lactones to polyesters and its mechanistic aspects, *International Journal of Biological Macromolecules* 25 (1999) 145.

[36] Uyama H., Kobayashi S.: Enzyme-catalysed polymerization to functional polymers, *J. Mol. Cat. B.* 19–20 (2002) 117.

[37] Marcilla R., de Geus M., Mecerreyes D., Duxbury C.J., Koning C.E., Heise A.: Enzymatic polyester synthesis in ionic liquids, *Eur. Polym. J.* 42 (2006) 1215.

[38] Fujimaki T.: Processability and properties of aliphatic polyesters, "BIONOLLE", synthesised by polycondensation reaction, *Polym.Deg. Stab.* 59 (1998) 209–214.

[39] Marten E., Müller R.-J., Deckwer W.-D.: Studies on the enzymatic hydrolysis of polyesters, *Polym. Deg. Stab.* 88 (2005) 371.

[40] Witt U., Müller R.-J., Deckwer W.-D.: Biodegradation behaviour and material properties of aliphatic/aromatic polyesters of commercial importance, *J. Eviron. Polym. Degrad.* 5 (1997) 81.

[41] Müller R.-J., Witt U., Rantze E., Deckwer W.-D.: Architecture of biodegradable copolyesters containing aromatic constituents, *Polym. Degr. Stab.* 59 (1998) 203.

[42] Stridsberg K.M., Ryner M., Albertsson A.-Ch.: Controlled ring-opening polymerization: polymers with designed architecture in: Degradable aliphatic polyesters, *Adv. Polym. Sci.* Vol. 157, Springer (2002).

[43] Edlund U., Albertsson A.-Ch.: Degradable polymer microspheres for controlled drug delivery in: Degradable aliphatic polyesters, *Adv. Polym. Sci.* Vol. 157, Springer (2002).

[44] Grigat E., Koch R., Timmermann R.: BAK 1095 and BAK 2195: completely biodegradable synthetic thermoplastics , *Polym. Deg. Stab.* 59 (1998) 223–226.

[45] Chiellini E., Corti A., D'Antone S., Solaro R.: Biodegradation of poly(vinyl alcohol) based materials, Prog. Polym. Sci. 28 (2003) 963.

[46] Bastioli C.: Properties and applications of Mater-Bi starch-based materials, *Polym. Deg. Stab.* 59 (1998) 263.

Chapter 3
Properties and applications

Chapter 3
Properties and applications

3.1. BIODEGRADABLE POLYMERS FROM RENEWABLE RESOURCES

3.1.1 Poly(lactic acid) – PLA

Properties

Poly(lactic acid) (PLA) exhibits a balance of performance properties that are comparable to those of traditional thermoplastics [1]. PLA can be fabricated in a variety of familiar processes and brings a new combination of attributes to packaging, including stiffness, clarity, deadfold and twist retention, low temperature heat sealability, as well as an interesting combination of barrier properties including flavour, aroma and grease resistance.

PLA polymers range from amorphous glassy polymers with a glass transition temperature of about 50–60°C to semicrystalline products with melting points ranging from 130 to 180°C, depending on the sequence of enantiomeric repeating units (L and D) in the polymer backbone. [2].

Generally, commercial PLA grades are copolymers of poly(L-lactic acid) (PLLA) and poly(D,L-lactic acid) (PDLLA), which are produced from L-lactides and D,L-lactides, respectively [3]. The ratio of L- to D,L-enantiomers is known to affect the properties of PLA, such as melting temperature and degree of crystallinity. Enantiometrically pure PLA, poly(L-lactide) is a semicrystalline polymer with a glass transition temperature (T_g) of about 55°C and melting point (T_m) of about 180°C [4–6]. Introduction of stereochemical defects into poly(L-lactide) (i.e. meso-lactide or D-lactide incorporation) reduces melting point, rate of crystallization, and extent of crystallization of the resulting polymer but has little effect on glass transition temperature [7]. After roughly 15% incorporation of mesolactide, the result is no longer crystallizable. For example, introduction of mesolactide depresses the crystalline melting point to 130°C [4].

The molecular weight, macromolecular structure and the degree of crystallization of PLA vary substantially depending on the reaction conditions in the polymerization process [8].

Of the three possible isomeric forms, poly(L-lactic acid) and poly(D-lactic acid) are both semicrystalline in nature, and poly(mesolactic acid) or poly(D,L-lactic acid) is amorphous. Racemic PLA – synthesized from petrochemicals – is atactic, i.e. it exhibits no stereochemical regularity of structure, is highly amorphous and has a low glass transition temperature. Amorphous grades of PLA are transparent.

PLA has good mechanical properties, thermal plasticity and biocompatibility, is readily fabricated, and is thus a promising polymer for various end-use applications. From a physical property standpoint it is often loosely compared to polystyrene [7]. Like polystyrene, standard-grade PLA has high modulus and strength and is lacking in toughness. The toughness of PLA can be dramatically improved through orientation, blending, or copolymerization [7].

Electrical properties of biodegradable polylactic acid films were measured and compared with those of crosslinked polyethylene (XLPE) currently used as insulation for cables and electric wire [10]. The volume resistivity, dielectric constant and dielectric loss tangent of PLLA were found to be almost the same as those of XLPE. However, the impulse breakdown strength of PLLA was 1.3 times that of XLPE.

Table 3.1. Properties of some commercially available PLA

	Nature Works® PLA [6]	Nature Works® PLA Resin General purpose [8]	Biomer® L9000 [6]	Hycail HM 1011 [9]
Physical properties				
Melt flow rate (g/10 min)		10–30	3–6	2–4
Density (g/cm^3)	1.25	1.24	1.25	1.24
Haze	2.2			
Yellowness index	20–60			
Clarity		Transparent		
Mechanical properties				
Tensile strength at yield (MPa)	53	48	70	62
Elongation at yield (%)	10–100	2.5	2.4	3–5
Flexural modulus (MPa)		3828	3600	
Flexural strength (MPa)		83		
Notched Izod impact (J/m)		0.16		
Thermal properties				
HDT (°C)	40–45, 135			
Vicat softening point (°C)	–*		56	
Glass transition temperature (°C)	55–65			60–63
Melting point (°C)	120–170**			150–175

* Close to glass transition temperature.
** Amorphous and crystalline, respectively.

Processing

PLA resin can be tailor-made for different fabrication processes, including injection moulding, sheet extrusion, blow moulding, thermoforming, film forming, or fibre spinning [7]. The key is controlling certain molecular parameters in the process, such as branching, D-isomer content, and molecular weight distribution. Injection moulding of heat-resistant PLA products requires rapid crystallization rates, which can be achieved by PLA that contains less than 1% D-isomer and often with the addition of nucleating agents [7]. Extrusion-thermoforming is optimized at a D-isomer content that does not allow crystallization to occur during the melt processing steps, with 4–8% D content being the effective range.

The recommended process temperature for Hycail PLA is 190–240°C [9].

The processing temperature profile of Nature Works PLA 3001 D polymer, designed for injection moulding applications, comprises: melt temperature 200°C, feed throat 25°C, feed temperature (crystalline pellets) 165°C, feed temperature (amorphous pellets) 150°C, compression section 195°C, metering section 205°C, nozzle 205°C and mould 25°C [8]. For extrusion grades the processing temperature profile ranges from 180 to 210°C.

Applications

Poly(lactic acid) products are finding uses in many applications, including packaging, paper coating, fibres, films, and a host of moulded articles.

The first products were aimed at packaging film and fibres for textiles and non-wovens. For packaging, it recommends clear films with good barrier properties but low heat-seal properties. For fibres it could mean apparel with better drape and moisture management and industrial clothing with better UV resistance, reduced flammability and good resistance to soiling and staining [11]. Cargill Dow's PLA has been designated as a new generic fibre type by the US Federal Trade Commission. PLA now joins other classifications including cotton, wool, silk, nylon, and polyesters as a recognized fibre category.

Examples of main applications for PLA are given in Table 3.2.

Table 3.2. Main applications for PLA

Sector	Examples
Packaging	Food packaging, films, rigid thermoformed food and beverage containers, carrier bags and labels, coated papers and boards, battery packaging, windows for envelopes
Agriculture	Sheet or moulded forms for time-release fertilizers, plant clips
Transportation	Parts of automobile interiors (head liners, upholstery, spare tyre covers)
Electric appliances and electronics	CD, computer keys, cases for Walkmans, wrappers for CD
Houseware	Carpets
Other (fibres and fabrics)	Textiles and non-wovens

In addition to traditional food packaging applications, several companies are exploring non-food packaging applications for PLA, including [1]:

* Mitsui-Chemical telephone cards
* Sanyo compact disc
* Matsushita (Panasonic) battery packaging
* Fujitsu PC body components

3.1.2 Polyhydroxyalkanoates – PHA

Properties

The family of polyhydroxyalkanoates (PHA) exhibits a wide variety of mechanical properties from hard crystalline to elastic, depending on the composition of monomer units [12]. Solid-state poly(3-hydroxybutyrate) (P(3HB)) is a compact right-handed helix with a two-fold screw axis (i.e. two monomer units complete one turn of the helix) and a fibre repeat of 0.596 nm [13]. The stereoregularity of P(3HB) makes it a highly crystalline material. Its melting point is around 177°C close to that of polypropylene, with which it has other similar properties, although the biopolymer is stiffer and more brittle.

The densities of crystalline and amorphous PHB are 1.26 and 1.18 g/cm^3, respectively [12].

P(3HB) is water insoluble and relatively resistant to hydrolytic degradation. This differentiates P(3HB) from most other currently available bio-based plastics which are either moisture or water soluble. Mechanical properties of PHB like Young's modulus and tensile strength are close to that of polypropylene though extension to break is markedly lower than that of polypropylene (Table 3.3) [12, 14]. However, due to the high stereoregularity of biologically produced macromolecules, PHB is a highly crystalline polymer that is stiff and brittle. It is also thermally unstable during processing [15]. The molecular weight of PHB degrades significantly at temperature just above the T_m. This unfortunate aspect of properties poses a

Table 3.3. Comparison of mechanical properties of PHAs and polypropylene [12, 14]

Polymer	Copolymer content	Melting temperature, °C	Young modulus, GPa	Tensile strength, MPa	Elongation at break, %
PP	–	170	1.7	34.5	400
P(3HB)	–	179	3.5	40	5
P(3HB-*co*-3HV)	3 mol% 3HV	170	2.9	38	–
P(3HB-*co*-3HV)	9 mol% 3HV	162	1.9	37	–
P(3HB-*co*-3HV)	14 mol% 3HV	150	1.5	35	–
P(3HB-*co*-3HV)	20 mol% 3HV	145	1.2	32	–
P(3HB-*co*-3HV)	25 mol% 3HV	137	0.7	30	–
P(3HB-*co*-4HB)	3 mol% 4HB	166	–	28	45
P(3HB-*co*-4HB)	10 mol% 4HB	159	–	24	242
P(3HB-*co*-4HB)	16 mol% 4HB	–	–	26	444
P(3HB-*co*-4HB)	64 mol% 4HB	50	30	17	591
P(3HB-*co*-4HB)	90 mol% 4HB	50	100	65	1080
P(4HB)	–	53	149	104	1000
P(3HHx-*co*-3HO)		61	–	10	300
P(3HB-*co*-6 mol% 3HA)		133	0.2	17	680
P(3HB-*co*-3HHx)		52	–	20	850

limitation of, for example, the application to a flexible film, which is one of the largest uses of biodegradable polymers. As a consequence, many attempts to copolymerize a comonomer with PHB monomer for improving its mechanical properties have been made. One idea is to include a more bulky comonomer to reduce the crystallinity and presumably increase the flexibility of the resulting copolymers. The copolymerization with 3-hydroxyvalerate (3-HV) was the first attempt performed by ICI in the early 1980s. However, the crystallinity of poly(3-hydroxybutyrate-*co*-3-hydroxyvalerate) (P(HB-*co*-HV)) never falls below 50% due to the isodimorphism of the P(HB-*co*- HV) copolymer. It has been reported that poly(3-hydroxybutyrate-*co*-3-hydroxyhexanoate) (P(HB-*co*HHx) shows a greater T_m, at a given mol% comonomer, in comparison to P(HB-*co*-HV). Interestingly, hexanoate and larger comonomers depress T_m in the same manner regardless of their molecular sizes. This feature indicates the breakdown of the isodimorphism occurring in the P(HB-*co*-HV) copolymer by the incorporation of comonomer units with three or more carbon units [15].

Incorporation of other hydroxy-acid units to form PHA copolymers can improve properties such as crystallinity, melting point, stiffness and toughness [12]. As the fraction of 3HV increases the copolymer becomes tougher (increase in impact strength) and more flexible (decrease in Young's modulus). The increase of melting temperature with increasing 3HV fraction without affecting degradation temperature allows thermal processing of copolymer melts without thermal degradation. The melting temperature (T_m) of P(3HB) homopolymer was 178°C, and the copolymer (3-hydroxybutyrate-*co*-3-hydroxyvalerate) (P(3HB-*co*-3HV)) with a 95 mol% of 3HV was 108°C [16]. A minimum value (around 75°C) of melting temperature was observed at approximately 40 mol% 3HV, where the crystal lattice transition took place. For copolymers poly(3-hydroxybutyrate-*co*-4-hydroxybutyrate)s (P(3HB-*co*-4HB)) the T_m value decreases from 178°C to 150°C as the 4HB content increases from 0 to 18 mol%, then is almost constant in the composition range from 18 to 49 mol% 4HB.

PHAs made of longer monomers, such as medium chain length mcl-PHAs, i.e. with C_6–C_{14} monomers, are typically elastomers and sticky materials, which can also be modified to make rubbers [17]. PHA copolymers composed of primarily HB with a fraction of longer chain monomers, such as HV, HH or HO, are more flexible and tougher plastics.

The copolymer P(3HB-*co*-3HV) has lower crystallinity and improved mechanical properties (decreased stiffness and brittleness, increased tensile strength and toughness) compared to P(3HB), while still being readily biodegradable. It also has a higher melt viscosity, which is a desirable property for extrusion blowing [6]. Copolymers PHBV poly(3-hydroxybutyrate-*co*-3-hydroxyhexanoate)s have a range of properties depending on composition (Table 3.4).

Table 3.4. Properties of poly(3-hydroxybutyrate-*co*-3-hydroxyhexanoate)s [18]

% C6 (hexanoate)	Melting point T_m, °C		Applications
0 (PHB)	180	Hard, brittle, crystalline	
4	150	Hard, some elasticity	moulded articles
6	145	Hard, elastic, flexible	fibres
10	125	Soft, elastic, flexible	films
18	95	Soft, rubbery	coatings

Poly(3-hydroxybutyrate-*co*-3-hydroxyhexanoate) combines the thermo-mechanical properties of PE (strength, flexibility, ductility, toughness, elasticity) with the physical–chemical properties (compatibility) of polyesters (printability, dyeability, barrier performance). It forms blends with PLA and thermoplastic starch.

Properties of some commercially PHAs are given in Table 3.5.

Table 3.5. Properties of commercially PHAs [6]

	P(3HB) Biomer P240	P(3HB) Biomer P226	P(3HB-*co*-3HV) Biopol	P(3HB-*co*-3HHx) Kaneka, Nodax
Physical properties				
Melt flow rate (g/10 min)	5–7	9–13		0.1–100
Density (g/cm^3)	1.17	1.25	1.23–1.26	1.07–1.25
Transparency (%)			0.7	White powder/ translucent film
Mechanical properties				
Tensile strength at yield (MPa)	18–20	24–27		10–20
Elongation at yield (%)	10–17	6–9		10–25
Flexural modulus (MPa)	1000–1200	1700–2000	40	Several orders of magnitude
Thermal properties				
HDT (°C)	–	–		60–100
Vicat softening point (°C)	53	96		60–120
Glass transition temperature (°C)				
Melting point (°C)				

Processing

Homopolymer P(3HB) has good thermoplastic properties (melting point 180°C) and can be processed as classic thermoplast and melt spun into fibres. It has a wide in-use temperature range (articles retain their original shape) from $-30°C$ to $120°C$. Articles made of P(3HB) can be autoclaved. However, it is fairly stiff and brittle, and has somewhat limiting applications. PHB has a small tendency to creep and exhibits shrinkage of 1.3%.

A comparison of injection moulding conditions between homopolymer PHB and PP is given in Table 3.6. [19].

Table 3.6. Injection moulding conditions of PHB and PP [19]

Parameters	PHB	PP
Melt temperature, °C	160	180
Hopper temperature, °C	25	25
Fill temperature, °C	130	230
Clamp zone, °C	140	250
Mixture zone, °C	150	250
Nozzle, °C	160	250
Mold, °C	10–15	10–15

PHBV is thermoplastic and can be processed by injection moulding, extrusion, blow moulding, film and fibre forming, and lamination techniques.

The Nodax family of PHAs are suitable for different conversion processes, including injection moulding, cast film, cast sheet for thermoforming, melt extruded paper and board coatings [20].

Applications

Initially, PHAs were used in packaging films mainly in bags, containers and paper coatings [12, 21]. Similar applications in conventional commodity plastics include disposable items such as razors, utensils, nappies, feminine hygiene products, cosmetic containers, shampoo bottles and cups [16]. P(3HB-*co*-3HHx) (Nodax) has applications in flushable materials (e.g. feminine hygiene products), coatings, synthetic papers, heat-formed products, binding materials, and films. Markets for the Nodax family of PHAs [20] include:

- Packaging
- Single use and disposable items
- Housewares
- Appliances
- Electrical and electronics
- Consumer durables
- Agriculture and soil stabilization
- Adhesive and soil stabilization
- Adhesives, paints and coatings
- Automotive

3.1.3 Thermoplastic starch – TPS

Properties

The glass transition temperature (T_g) of dry amorphous starch is experimentally inaccessible owing to the thermal degradation of starch polymers at elevated temperatures [22]. It is estimated the T_g of the dry starch to be in the range of 240–250°C [22]. Native starch is a non-plasticized material because of the intra- and intermolecular hydrogen bonds between hydroxyl groups of starch molecules. During the thermoplastic process, in the presence of a plasticizer, a semicrystalline granule of starch is transformed into a homogeneous material with hydrogen-bond cleavage between starch molecules, leading to loss of crystallinity.

The physical properties of the thermoplastic starch are greatly influenced by the amount of plasticizer present. In most literature for thermoplastic starch, polyols were usually used as plasticizers, of which glycerol is the major one. The effect of plastification level on glass transition of thermoplastic starch is presented in Table 3.7.

Table 3.7. Glass transition of TPS using different plastification levels [23]

% starch	Plasticizer level, wt%	Glycerol content, wt%	Water content, wt%	Glass transition, °C
74	26	10	16	43
70	30	18	12	8
67	33	24	9	−7
65	35	35	0	−20

According to the plasticizer/starch, thermoplastic starch presents a large range of properties. A number of studies on the effects of plasticizers on properties of thermoplastic starch have been carried out. Plasticizers used include polyols such as glycerol, glycol, xylitol, sorbitol, and sugars and etanoloamine [24–32]. Plasticizers containing amide groups such as urea, formamide and acetamide or a mixture of plasticizers have been also studied [33–37].

The mechanical properties of a low and a high molecular mass thermoplastic starch were monitored at water contents in the range of 5–30% (w/w). The stress–strain properties of the materials were dependent on the water content. Materials containing less than 9% water were glassy with an elastic modulus between 400 and 1000 MPa [24]. Different starch sources were extruded with the plasticizer glycerol and glass transition temperatures and mechanical properties were evaluated [24]. Above certain glycerol contents, dependent on the starch source, a lower glass transition temperature T_g resulted in decreased modulus and tensile strengths and increased elongations. For pea, wheat, potato and waxy maize starch the T_g was 75°C, 143°C, 152°C and 158°C, respectively.

The effect of the type and amount of plasticizer on the mechanical, thermal and water-absorption properties of melt-processed starch was investigated [31]. It was reported that, in general, monohydroxyl alcohols and high molecular weight glycols failed to plasticize starch, whereas shorter glycols were effective.

The mechanical properties of starch-based plastics of native corn, potato, waxy corn and wheat starch, produced by compression moulding of native starch and glycerol in the weight ratio 0 to 3 were strongly dependent on the water content and starch source [38].

The mechanical and melt flow properties of two thermoplastic potato starch materials with different amylose contents were evaluated [32]. After conditioning at 53% relative humidity

(RH) and 23°C, the glycerol-plasticized sheets with a high amylose content (HAP) were stronger and stiffer than the normal thermoplastic starch with an amylose content typical for common potato starch. The tensile modulus at 53% RH was about 160 MPa for the high amylose material and about 120 MPa for the plasticized NPS (native potato starch). The strain at break was about 50% for both materials.

Table 3.8. Properties of thermoplastic starches

	Potato thermoplastic starch [38]	Wheat thermoplastic starch [39]*
Physical properties		
Melt flow rate (g/10 min)		
Density (g/cm^3)		1.34–1.39
Transparency (%)		
Mechanical properties		
Tensile strength at yield (MPa)	22	1.4–21.4
Elongation at yield (%)	3	3–104
Tensile modulus (MPa)	1020	11–1144
Thermal properties		
Glass transition temperature (°C)		(−)20–43
α-transition (DMTA) (°C)		1–63

* Properties after equilibrium at 23°C and 50%, 6 weeks; glycerol to starch ratio: 0.135–0.538; water content: 16–0 wt%.

Processing

Various industrial processing techniques have been used to prepare starch plastics, including kneading, extrusion, compression moulding and injection moulding [40]. Processing temperatures are in the range of 100–200°C, although care has to be taken at temperature above 175°C because of starch molecular breakdown [40–42]. Most research has been focused on water and glycerol as the most important additives. As melt flow accelerators lection, glycerin monostearate and calcium stearate have been studied. Several native starches have been processed, such as: wheat, rice, corn, waxy maize starch, high amylose corn starch and potato starch [40]. The dimensions of moulded objects from hydrophilic polymers such as starch depend on their water content [43]. If precise dimensions are required, processing should be carried out so that products are formed at approximately the equilibrium in-use water content. For potato starch, for example, this means water contents of around 14% for use under ambient conditions (50% relative humidity, 20–25°C) [43]. If higher water contents are used in processing, distortion and shrinkage will occur as the equilibrium water content is naturally achieved after processing. In addition, higher water content can induce more hydrolytic degradation of the starch chains during processing and also gelatinization rather than melt formation. If lower water contents are used, thermal degradation can occur during processing, as well as swelling after processing.

Applications

The first commercial product made of injection-moulded thermoplastic starch was the drug-delivery capsule Capill, and further products are gradually appearing, e.g. golf tees, cutlery, plates, and food containers [43]. In addition, extrusion has been applied to produce rigid foams, suitable for loose-fill packaging.

Packaging is the dominant application area for starch-based polymers [6]. Main application areas include:

- foams (for the loose-fill foam market)
- films (for agriculture, e.g. mulch films)
- shopping bags
- mouldable products (pots, cutlery, fast food packaging)

3.1.4 Other compostable polymers from renewable resources

Cellulose

Properties
Cellulose esters were, besides cellulose esters of inorganic esters and cellulose ethers, pioneer compounds of cellulose chemistry, and remain the most important technical derivatives of cellulose [44]. Unlike commodity plastics such as polyolefins, cellulose cannot be processed thermoplastically. However, derivatization, i.e. esterification, can yield materials suited for thermoplastic processing. Cellulose esters, such as cellulose acetate (CA), cellulose acetate propionate (CAP), and cellulose acetate butyrate (CAB), are thermoplastic materials produced through the esterification of cellulose [45]. A variety of raw materials such as cotton, recycled paper, wood cellulose, and sugarcane are used in making cellulose ester biopolymers in powder form. Such powders combined with plasticizers and additives are extruded to produce various grades of commercial cellulosic plastics in pelletized form. Of great interest as potential biodegradable plastics are also long-chain aliphatic acid esters of cellulose [46, 47].

Cellulose esters characterize stiffness, moderate heat resistance, high moisture vapor transmission, grease resistance, clarity and appearance, and moderate impact resistance [47].

Some properties of commercial cellulose esters are given in Table 3.9.

Table 3.9. Properties of cellulose esters [48]

	Cellulose acetate propionate Albis CAP CP800 (10% plasticizer)	Cellulose acetate butyrate Albis CAB B900 (10% plasticizer)
Physical properties		
Melt flow rate (g/10 min)		
Density (g/cm^3)	1.21	1.19
Water absorption at 24 hrs	1.6	1.4
Mechanical properties		
Tensile strength at yield (MPa)	31.7	28.3
Elongation at break (%)	30	30
Flexural modulus (MPa)	1240	1170
Flexural strength (MPa)	41.4	37.2
Thermal properties		
HDT (°C)		
Vicat softening point (°C)	102	104
GTT (°C)		
Melting point (°C)		

Processing
Cellulose esters are easy materials to extrude and injection mould [47]. Some of the innate properties include a relatively narrow window between the melt flow temperature and the decomposition temperature. Therefore, in most commercial applications, plasticizers are used in conjunction with cellulose esters. Triethyl citrate is usually used for cellulose acetate (CA) and dioctyl adipate for cellullose acetate propionate (CAP).

Through plasticization of cellulose acetate by an environmentally friendly triethyl citrate plasticizer, the cellulose acetates are processable at 170°C–180°C, much below the melting point of cellulose acetate (233°C) [45]. Materials processed by extrusion followed by injection moulding exhibited better properties as compared to those processed by extrusion followed by compression moulding, as additional shear forces applied during injection moulding resulted in stiffer product. Cellulosic plastics fabricated through injection moulding at a higher temperature (190°C) exhibited better tensile properties over their counterparts injected moulded at a comparatively lower temperature (180°C) [45].

Applications
Materials such as metal, plastic, wood, paper, and leather are coated with polymers primarily for protection and for the improvement of their properties. For this purpose, cellulose acetate (CA), cellulose acetate propionate (CAP), and cellulose acetate butyrate (CAB) are the most important classical and solvent-based cellulose esters of the coating industry [44]. Cellulose esters are widely used in composites and laminates as binder, filler, and laminate layers. In combination with natural fibres, they can be used to some extent as composites from sustainable raw materials with good biodegradability. An additional domain of cellulose esters is their use in controlled-release systems, as well as membranes and other separation media [44, 47]. In the field of controlled-release systems, cellulose esters are used as enteric coatings, hydrophobic matrices, and semipermeable membranes for applications in pharmacy, agriculture, and cosmetics.

Other applications of cellulose esters include:

- thin films
- containers
- handles
- optical applications
- automotive applications
- toys
- writing instruments
- electric insulation films, lights and casings

Chitosan

Properties
Chitin and chitosan are examples of highly basic polysaccharides. Chitin is highly hydrophobic and is insoluble in water and most organic solvents. It is a hard, white, inelastic, nitrogenous polysaccharide [49]. An important parameter, which influences its physical–chemical and biomedical characteristics, is the degree of N-acetylation, especially in chitosan. Converting chitin into chitosan lowers the molecular weight, changes the degree of N-acetylation, and thereby alters the net charge distribution, which in turn influences the degree of agglomeration [49]. The average molecular weight of chitin is 1.03 to 2.5×10^6 Da, but upon N-deacetylation, it

reduces to 1.0 to 5×10^5. Chitosan is soluble in dilute acids such as acetic acid, formic acid, etc. Chitosan has many useful characteristics such as hydrophilicity, biocompatibility, biodegradability, and antibacterial characteristics [49–51]. Chitin and chitosan degrade before melting, which is typical of polysaccharides with extensive hydrogen bonding [49].

Chitosan can form transparent film, which may find application in a variety of packaging needs [49, 52]. In 1936, Rigby was granted a patent for making film from chitosan and a second patent on making fibres from chitosan [49]. The films were described as flexible, tough, transparent, and colourless with a tensile strength of about 6210 kPa.

Plasticizing agents are essential generally to overcome the brittleness of the biopolymeric films. Chitosan films were prepared by blending with polyols (glycerol, sorbitol and polyethylene glycol (PEG)) and fatty acids (stearic and palmitic acids) and their mechanical and barrier properties were studied [52]. The tensile strength of the blended films decreased with the addition of polyols and fatty acids, whereas the percent elongation was increased in polyol blend film, but fatty acid blend films showed no significant differences. Glycerol blend film showed a decrease, whereas sorbitol and PEG blend film showed an increase in water vapour permeability (WVP) values.

Processing
Chitosan possesses an excellent ability to form porous structures [50]. It can be moulded in various forms as porous membranes, blocks, tubes and beads. Chitosan also readily forms films and produces material with very high gas barrier. Chitosan films are prepared by dissolving chitosan in dilute acid and spreading on a levelled surface and air-drying at room temperature [49]. Films are also prepared by drying at 60°C in an oven by spreading the solution on plexiglass.

Applications
Chitosan has prospective applications in many fields such as biomedicine, waste water treatment, functional membranes and flocculation [49, 53, 54]. Chitosan has been used in the purification of drinking water and in cosmetics and personal care products. Due to its excellent biological properties such as biodegradation in the human body, biocompatibility, and immunological, anti-bacterial, and wound-healing activities it has also a variety of medical uses such as wound dressings, drug delivery, encapsulation, etc. [54]. Chitosan has found a potential application as a support material for gene delivery, cell culture and tissue engineering. It is also known as an adsorptive material, e.g. sorbent for heavy metal ions. It has been used for the production of edible coatings. Chitosan films were used in extending the shelf life of vegetables [49].

Proteins

Properties
Until recently, the only uses and applications for proteins were in food sciences [55]. The development of studies on non-food uses of agricultural raw materials initiated an interest in protein-based plastics. A number of proteins of plant origin have received attention for the production of biodegradable polymers. These proteins are corn zein, wheat gluten, soy protein, and sunflower protein.

The major drawback of protein-based plastics, apart from keratin, is their sensitivity towards relative humidity. For example, it was reported that after being submerged in water for 20 h the soy protein sheets absorbed up to 180% water [56].

Soy protein plastics are rigid, but tend to be brittle and water sensitive [56, 57]. Water resistance of soybean protein-based plastics can be improved by chemical modification of the proteins, or blending, e.g., with polyesters [57]. The flexibility of soybean protein-based plastics can be improved by adding various plasticizers [56, 57]. It was reported that depending on the moisture and glycerol contents, soy protein plastic sheets displayed properties from rigid to soft [56]. The glass transition temperature of the sheets varied from ca. -7 to 50°C with moisture contents ranging from 26 to 2.8% and 30 parts of glycerol.

Among proteins, wheat gluten with its unique viscoelastic properties and its water insolubility is of particular interest for the preparation of biodegradable polymer materials. To control the brittleness of protein-based materials and to lower their shaping temperature, the addition of plasticizer is generally required [58]. Water and glycerol are common plasticizers of wheat gluten. Other compounds including polyols, sugars, ethylene glycol and its derivatives, lipids and emulsifiers have been tested as gluten plasticizers. Various compounds, differing in their chemical functions, number of functional group and degree of hydrophobicity, including water, glycerol, 1,4-butanediol, lactic and octanoic acids, were tested as wheat gluten plasticizers in a thermoplastic process [58].

The glass transition temperature (T_g) of hydrophobized and native wheat gluten and its protein fractions, with water mass fraction from 0 to 0.2, was studied using modulated differential scanning calorimetry [57]. The T_g values of unplasticized products were approx. 175°C whatever the treatment (hydrophobization) or the fraction tested, except for the gliadin-rich fraction (162°C) [59]. Thermal properties of corn gluten meal and its proteic components were investigated by Di Giola *et al.* (Table 3.10) [60].

Processing, and modification routes to produce and improve properties of biodegradable plastics from soy isolate were studied [62]. Soy isolate, acid-treated and crosslinked soy were subsequently compounded, extruded, and injection moulded. The obtained plastics were rigid and brittle with stiffness ranging from 1436 MPa for soy, to 1229 MPa for glyoxal crosslinked soy, up to 2698 MPa for heat-treated soy.

Table 3.10. Glass transitions temperatures of protein materials [60]

Material	Glass transition temperature, °C	Conditions	Technique
Corn gluten material	176	0% moisture	DMTA; MDSC
Extracted proteic component of corn gluten (zein)	164	0% moisture	DMTA; MDSC
Extracted proteic component of corn gluten (glutelin)	209	0% moisture	DMTA; MDSC

The influence of a set of hydrophilic plasticizers varying in their chain length (ethylene glycol and longer molecules) on the tensile strength and elongation at break of cast gluten films was studied [64]. Properties of deamidated gluten films enzymatically crosslinked were studied [65]. The action of transglutaminase with or without the addition of external diamines induced a simultaneous increase in tensile strength and elongation at break but tended to decrease the contact angle between the film surface and a water droplet.

Table 3.11. Mechanical properties of soy protein sheets [61]

Glycerol, parts	Stress at yield point, MPa	Elongation at yield point, %	Tensile strength, MPa	Young's modulus, MPa	Toughness, MPa
10	40.6	2.4	40.6	1226	0.4
20	33.9	7.9	34.0	1119	21.2
30	15.0	8.8	15.6	374	18.8
40	1.6	2.5	9.1	176	13.0
50	1.5	4.3	7.1	144	11.1

Table 3.12. Mechanical properties of wheat gluten materials plasticized with different amounts of water [63]

Sample	Water content, %	Tensile strength, MPa	Elongation at break, %	Young's modulus, MPa
W1	13.8	13.6	19.2	219.3
W2	15.7	7.5	57.4	143.0
W3	18.8	4.9	79.2	104.8
W4	21.2	3.0	91.4	67.5
W5	24.2	2.3	84.3	77.8

The effect of various crosslinked or hydrophobic additives (aldehydes, plant tannins, alcohols and fatty acids) on mechanical properties and water resistance of thermo-moulded films made from a sunflower protein isolate plasticized with glycerol were studied [66]. The use of octanoic acid resulted in high tensile strength (7 MPa), whereas the use of octanol resulted in a great increase in tensile elongation (54%). Several polyalcohols (glycerol, ethylene glycol, diethylene glycol, triethylene glycol and propylene glycol) were tested as sunflower proteins plasticizers [67]. The additives produced soft, brown and smooth films with good mechanical properties (σ_{max} = 6.2–9.6 MPa); ε_{max} = 23–140%) with a high level of impermeability to water vapour (1.9–9.9 $\times$ 10^{-2} g m^{-1} s^{-1} Pa^{-1}). However, these films were only moderately resistant to water. Glycerol and triethylene glycol were proposed as the most suitable plasticizers for sunflower proteins.

With a worldwide production estimated at about 33 million metric tonnes, cottonseed is the most important source of plant proteins after soybeans [68]. The viscoelastic behaviour of cottonseed protein isolate, plasticized with glycerol, was characterized in order to determine the temperature range within which cottonseed protein-based materials can be formed by extrusion or thermo-moulding [68]. The results indicated that cottonseed proteins are thermoplastics with a T_g ranging from 80 to 200°C when the glycerol content varies from 0% to 40% (w/w, dry basis).

Processing
Two important processes are used to make protein-based films: a wet process based on dispersion or solubilization of proteins, and a dry process based on thermoplastic properties of proteins under low water conditions [55].

Effects of moulding temperature and pressure on properties on soy protein polymers were evaluated [69]. The maximum stress of 42.9 MPa and maximum strain of 4.61% of the specimen were obtained when soy protein isolate was moulded at 150°C. Native soy protein was converted into a thermoplastic material in a corotating twin-screw extruder in the presence of 35% water and 10% glycerol (w/w relative to the protein amount) [62]. The extrusion was carried out at temperatures ranging from 70 to 80°C (temperature necessary for the splitting of the disulphide bridges and loss of the tertiary structure of the protein). Glycerol plasticized wheat gluten sheet was produced by extrusion at the barrel and die set temperature of 130°C [70].

Applications
Protein-based plastics have been used, alone or in mixtures, to obtain edible films and coatings. They have been used to protect pharmaceuticals and to improve the shelf life of food products. Some commercialization of protein films has been realized in collagen sausage casing, gelatin pharmaceutical capsules and corn zein protective coatings for nutmeats and candies [71].

Soybean protein can be used to produce a wide variety of non-food products, including plastic films, building composites, insulating foams, plywood adhesives, and other wood bonding agents [72].

The thermoplasticity and good-film forming properties of wheat gluten may be used to produce natural adhesives [73]. Gluten's adhesive properties make it useful in pressure-sensitive medical bandages and adhesive tapes. Gluten has the ability to provide edible protection for food or food components from interactions with the environment as they can serve as barriers to mass transfer (e.g. oxygen, water vapour, moisture, aroma, lipids) [73].

Some examples of application of proteins from various sources are given in Table 3.13.

Table 3.13. Examples of proteins technical applications [74]

Protein	Technical application
Soybean protein	Paper coatings, plywood adhesives
Maize zein	Printing inks, floor coatings, grease-proof paper
Keratin	Textiles, cosmetics
Rapeseed meal protein	Adhesives, plastics
Wheat gluten	Adhesives, coatings, cosmetics

3.2. BIODEGRADABLE POLYMERS FROM PETROCHEMICAL SOURCES

3.2.1 Aliphatic polyesters and copolyesters

Properties
Poly(butylene succinate) (PBS) is a commercially available, aliphatic polyester with many interesting properties, including biodegradability; melt processability, and thermal and chemical resistance [75–77]. PBS has excellent processability, so can be processed in the field of textiles into melt blow, multifilament, monofilament, flat, and split yarn and also in the field of plastics into injection moulded products, thus being a promising polymer for various potential applications [75].

Commercial aliphatic polyesters and copolyesters under the tradename Bionolle (Showa Highpolymer, Japan) are white crystalline thermoplastics, have melting points ranging from about 90 to 120°C, glass transition temperatures ranging from about -45 to -10°C and density of about 1.25 g/cm^3. The main physical and mechanical properties of various Bionolle grades, including 1000 series (poly(butylene succinate)), 2000 and 3000 series (poly(butylene succinate adipate)), and 6000 series (poly(ethylene succinate)) are given in Table 3.14.

Table 3.14. Properties of typical grades of Bionolle [75]

Property	PBSU #1000	PBSU #2000	PBSU #3000	PESU #6000	LDPE F082	HDPE 5110	PP 210
MFR$^{190°C}$ (g/10 min)	1.5	4.0	28	3.5	0.8	11	3.0*
Density (g/cm^3)	1.26	1.25	1.23	1.32	0.92	0.95	0.90
Melting point (°C)	114	104	96	104	110	129	163
Glass transition temperature (°C)	-32	-39	-45	-10	-120	-120	-5
Yield strength (kg/cm^2)	336	270	192	209	100	285	330
Elongation (%)	560	710	807	200	700	300	415
Stiffness 10^3 (kg/cm^3)	5.6	4.2	3.3	5.9	1.8	12.0	13.5
Izod impact strength** (kg-cm/cm) 20°C	30	36	>40	10	>40	4	2
Combustion heat (cal/g)	5550	5640	5720	4490	>11 000	>11 000	>11 000

* MFR was measured at 230°C.
** Izod impact strength was measured with notched samples.

Effects of ethyl and n-octyl branches on the properties of poly(butylene succinate) (PBS) and poly(ethylene adipate) (PEA) were investigated [78]. Glass transition and melting temperature, crystallinity, melt viscosity, and spherulite growth rate were decreased with an increase in the degree of the chain branches. The addition of n-octyl branches improved elongation and tear strength of PBS considerably without a noticeable decrease of tensile strength and modulus. The influence of polyester composition on thermal and mechanical properties of a series of aliphatic homopolyesters and copolyesters prepared from 1,4-butanediol and dimethyl esters of succinic and adipic acids was studied by Tserki *et al.* [79, 80]. The homopolymer poly(butylene succinate) is a highly crystalline polymer exhibiting a melting point (T_m) of 114.1°C and heat of fusion (ΔH_f) of 68.4 J/g, while for poly(ethylene adipate), which is a less crystalline polymer, the corresponding values are 60.5°C and 52.8 J/g, respectively. Copolyesters exhibited an intermediate behaviour depending on their composition. Glass transition temperature T_g decreased with increasing adipate unit content from -31.3 to -60.7°C. The homopolymer poly(butylene succinate) exhibited the highest tensile strength, which decreased with increasing adipate unit content, passed through a minimum at copolyester close to equimolarity and then increased towards the value of poly(ethylene adipate). It was observed that in contrast to tensile strength, the elongation at break increased for adipate unit content of 20–40 mol%. Chain extension reaction resulted in increase of polyester molecular weight leading to increased tensile strength [80]. Polyester crystallinity and melting temperature decreased upon chain extension, while glass transition temperature increased.

Crystallization and melting behaviour of polyesters based on succinic acid and respective aliphatic diols, with 2–4 methylene groups were studied by Papageorgiou *et al.* [80]. The equilibrium melting points were found to be 114, 133.5 and 58°C for poly(ethylene succinate), poly(butylene succinate) and poly(propylene succinate), respectively. The corresponding values for enthalpy of fusion were 180, 210 and 140 J/g. Poly(propylene succinate) exhibited the slowest crystallinization rates and lowest degree of crystallinity among these polyesters.

Processing
PBS may be processed using conventional polyolefin equipment in the range 160–200°C [75]. Injection, extrusion or blow moulding is suitable for processing PBS.

Applications
Applications include mulch film, cutlery, containers, packaging film, bags and "flushable" hygiene products [75].

3.2.2 Aromatic polyesters and copolyesters

Properties
As an engineering thermoplastic, poly(trimethylene terephthalate) (PTT) has a very desirable property set, combining the rigidity, strength and heat resistance of poly(ethylene terephthalate) (PET) with the good processability of poly(butylene terephthalate) (PBT) [6].

PTT is crystalline, hard, strong and extremely tough. The density of PTT is slightly lower than PET and similar to PBT. The tensile strength and flexural modulus decrease between PET, PTT and PBT, respectively (Table 3.15). The thermal and relaxation characteristics of PTT are intermediate to the properties of PET and PBT, and are typical of those encountered with semiflexible polymers of low to medium crystallinity [82]. The reported equilibrium melting temperature for PTT is approximately 237°C, with a corresponding 100% crystalline heat of fusion estimated to be 30 kJ/mol [83].

To improve the thermal and mechanical properties of biodegradable aliphatic polyesters, introducing aromatic terephthalate units into the main chain of aliphatic polyesters has been

Table 3.15. Properties of poly(trimethylene terephthalate) [6, 48]

	PTT [6]	PTT RTP 4700 [48]	PET [6]	PBT Celanex 1300A [48]
Physical properties				
Melt flow rate (g/10 min)				90
Density (g/cm^3)	1.35	1.33	1.40	1.31
Haze (%)			2–3	
Mechanical properties				
Tensile strength at yield (MPa)	67.6	61	72.5	55.2
Elongation at yield (%)		>10%		
Tensile modulus (MPa)		2551		
Flexural strength (MPa)		98		82.7
Flexural modulus (MPa)	2760	2758	3110	2200
Thermal properties				
HDT (°C)	59		65	
Vicat softening point (°C)			265	
GTT (°C)	45–75		80	60
Melting point (°C)	225			225

considered to produce aliphatic–aromatic copolyesters with better physical properties as well as still having biodegradability [84].

The solid-state microstructures and thermal properties of aliphatic–aromatic copolyesters of poly(butylene adipate-*co*-butylene terephthalate) were investigated by wide-angle X-ray (WAXD), solid-state ^{13}C NMR, differential scanning calorimetry (DSC) and atomic force microscopy (AFM) [84]. Both the melting temperature and crystallinity of copolyesters showed minimum values at around 25 mol% butylene terephthalate content, which is the transition point from PBA crystal structure to PBT crystal structure. It was reported that introducing 40 mol% or more butylene adipate units could reduce the glass transition temperature T_g of the copolyesters from 66°C to below −10°C, and reduce the melting temperature T_m from above 200°C to about 100°C [84]. Biodegradable ideal random copolymer poly(butylene adipate-*co*-terephthalate) (PBAT) was melt-spun into fibres with a take-up velocity up to 5 km/min [85]. Despite the ideal randomness and composition (1:1) of PBAT copolymers, PBAT fibre showed a well-developed PBT-like crystal structure, while its melting temperature (ca. 121°C) was over 100°C lower than that of PBT.

Ecoflex (poly(butylene adipate-*co*-terephthalate)), a commercialized aliphatic–aromatic copolyester from BASF, was characterized to be an ideal random copolymer with 44 mol% of BT units. The glass transition occurs at −30°C, and the melting point is 110–115°C [86]. The physical and mechanical properties of this soft thermoplastic are similar to those of LDPE, and it can be processed on conventional equipment for LDPE.

Processing

For injection moulding processing of RTP poly(trimethylene terephthalate) melt temperature and mould temperature were suggested to be between 232–260°C and 88–121°C, respectively [48]. Poly(trimethylene terephthalate) (PTT) can be spun and drawn at high speeds, resulting in a fibre suitable for applications such as sportswear, active wear, and other specialty textiles [6]. Poly(trimethylene terephthalate) has been melt spun at various take-up velocities from 0.5 to 8 km/min to prepare fibre samples [87]. The effect of take-up velocity on the structure and properties of as-spun fibres has been characterized through measurements of fibre fringence, density, wide-angle X-ray scattering, DSC melting behaviour, tensile properties and boiling water shrinkage.

The processing temperature of Ecoflex copolymer is 140–170°C (melt temperature) [48].

Applications

Poly(trimethylene terephthalate) (PTT) is an opaque rigid thermoplastic useful for many structural applications, e.g. in carpet, textile, film and packing and other engineering thermoplastic markets, where rigidity, strength and toughness are required [88].

PTT may be used to produce fibres for carpets and industrial textiles where it has good resiliency and wearability of nylon, combined with the dyeability, static resistance and chemical resistance of PET [6]. As a spunbond fibre for apparel, its property set includes good stretch recovery, softness and dyeability.

Main applications include:

- Fibres (textile, carpet, apparel)
- Packaging (films)

According to the manufacturer (BASF) Ecoflex has been developed for the flexible films sector. Typical applications include agricultural films, carrier bags and compost bags. The material is marketed as a compostable packaging film, as a hydrophobic protective coating for food containers, and as a blend component [86]. Copolyesters with a higher terephthalic acid unit have been reported to be suited for fibre applications.

Table 3.16. Properties of aliphatic–aromatic copolyesters [48]

	Ecoflex® F* [48]
Physical properties	
Melt volume flow rate (cm³/10 min)	3.5
Density (g/cm³)	1.26
Transmittance (%)	82
Mechanical properties	
Tensile strength at yield (MPa)	35–44
Tensile strength at break (%)	560–710
Tensile modulus (MPa)	
Flexural strength (MPa)	
Flexural modulus (MPa)	
Shore D hardness	332
Thermal properties	
HDT (°C)	
Vicat softening point (°C)	80
GTT (°C)	
Melting point (°C)	112

* A copolyester mainly based on 1,4-butanediol, adipic acid and terephthalic acid.

3.2.3 Poly(caprolactone) – PCL

Properties
Polycaprolactone (PCL) was developed as a biodegradable plastic of aliphatic polyester type derived from the chemical synthesis of crude petroleum [89]. It has a low melting point (ca. 60°C), low viscosity of its melt and it is easy to process [89]. PCL has good water, oil and chlorine resistance.

The PCL chain is flexible and exhibits high elongation at break and low modulus. The elongation at break and tensile strength of PCL films have been reported to be between 450 and 1100% and 25 and 33 MPa, respectively [90, 91]. These values are quite high as compared with the elongation at break, 500–725%, and tensile strength, 9.7–17.2 MPa, of low density polyethylene [91]. The main drawback of PCL is its low melting point which can be overcome by blending it with other polymers or by radiation crosslinking processes resulting in enhanced properties for a wide range of applications [77].

Properties of commercially available CAPA and Tone polycaprolactones are given in Tables 3.17 and 3.18 [92, 93].

Processing
PCL can be processed by the usual thermoplastic processing techniques, including blows and slot cast film extrusion, sheet extrusion, and injection moulding. The low melting point of PCL polymers requires lower temperatures than polyethylene and other polyolefins.

Table 3.17. Properties of CAPA polycaprolactones [92]

Property	CAPA 6500	CAPA 6800
Molecular weight, Mn	47 50 ± 2000	69 000 ± 1500
Melting point, °C	60–62	60–62
Heat of fusion, J/g	76.9	76.6
Crystallinity, %	56	56
Crystallization temperature, °C	25.2	27.4
Glass transition, °C	−60	−60
Melt flow rate (MFR), g/10 min (190°C/2.16 kg)	28	7.29
Tensile yield stress, MPa	17.2	14
Tensile modulus, MPa	430	500
Strain at break, %	>700	920
Flexural modulus, MPa	411	nd
Hardness, Shore D	51	50
Viscosity, Pa.sec (70°C, 10 1/sec)	2890	12 650
Viscosity, Pa.sec (100°C, 10 1/sec)	1353	5780
Viscosity, Pa.sec (150°C, 10 1/sec)	443	1925

Table 3.18. Properties of Tone polycaprolactones [93]

Property	P-767	P-787
Density, g/cm^3	1.145	1.145
Melt rate (MFR), g/10 min (190°C)	30	4
Tensile strength, MPa	21.3	39.7
Tensile modulus, MPa	435	386
Ultimate elongation, %	600–800	750–900
Flexural modulus, MPa	575	514
Flexural stress at 5% strain, MPa	23.4	21.0
Izod impact, J/m (notched)	82	350
Izod impact, J/m (unnotched)	No break	No break
Water absorption	0.3508	0.3295
Hardness, Shore D	55	55

According to manufacturers' information the extrusion parameters for polycaprolactones are: 70–120°C (CAPA 6500) and 130–165°C (CAPA 6800) [92].

Applications

PCL was recognized as a biodegradable and non-toxic material. Its high permeability to low molecular species at body temperature and biocompatibility makes PCL a promising candidate for biomedical applications, such as controlled drug delivery [94]. PCL is used mainly in thermoplastic polyurethanes, resins for surface coatings, adhesives and synthetic leather and

fabrics [89]. It also serves to make stiffeners for shoes and orthopaedic splints, and fully biodegradable bags, sutures, and fibres [89]. PCL is often mixed with starch to obtain a good,
biodegradable, low cost material.

The main applications of PCL comprise [92]:

- biodegradable bottles
- biodegradable films
- controlled release of drugs, pesticides and fertilizers
- polymer processing
- adhesives
- non-woven fabrics
- synthethic wound dressings
- orthopaedic casts

3.2.4 Poly(esteramide)s – PEA

Properties
Poly(esteramide)s constitute a new series of thermoplastic polymers that can combine high
technical performance with good biodegradability [95–99]. BAK 1095 is an example of a
poly(esteramide) that has been recently commercialized by Bayer. This is a statistical polymer with an amide/ester ratio of 6/4 based on 1,4-butanediol, adipic acid and 1,6 aminohexa-
noic acid. BAK poly(esteramide)s differing in the amide/ester ratio have been synthesized and
characterized [96]. Spectroscopic analyses of BAK poly(esteramide)s with 50/50, 60/40 and
70/30 amid/ester ratios showed a random distribution of monomers, which was in agreement
with their low crystallinity (12–14%). BAK polymers showed a decrease in the melting and
glass transition temperatures when the ester/amide/ratio was increased. In the same way,
Young's modulus decreased (Table 3.19). Influence of substitution of adipic acid by terephthalic acid units on thermal and mechanical properties of poly(esteramide)s was investigated
[97]. A regular increase in glass transition and melting temperatures with the aromatic content
was observed. Moreover, the mechanical properties showed an increase in chain stiffness with
the aromatic content.

Table 3.19. Properties of poly(esteramide)s [96, 99]

Property	BAK 1095 [99]	BAK 70/30 [96]	BAK 60/40 [96]	BAK 70/30 [96]
Density, g/cm^3				
Melt flow rate (MFR), g/10 min (190°C)				
Tensile strength, MPa	27	29	27	11
Tensile modulus, MPa	250	285	250	128
Elongation at break, %	570	432	570	24

High molecular weight segmented poly(esteramide)s comprising different ester to amide
ratios have been prepared by melt polycondensation of a preformed bisamide-diol, 1,4-
butanediol and dimethyl adipate [98]. The polymers had a low and a high melt transition, corresponding with the melting of crystals comprising single esteramide sequences and two or
more esteramide sequences, respectively. The low melt transition is between 58 and 70°C and

is independent of polymer composition. By increasing the hard segment content from 10 to 85 mol% the high melt transition increased from 83 to 140°C while the glass transition temperature increased from −45 to −5°C. Likewise, the elastic modulus increased from 70 to 524 MPa, the stress at break increased from 8 to 28 MPa while the strain at break decreased from 820 to 370%.

Processing

The processing conditions of BAK poly(esteramide) are similar to those of polyolefins [99]. BAK 1095 resin can be processed into film and also into extruded or blow-moulded parts on conventional machinery used for processing thermoplastics. Processing conditions are given in Table 3.20.

Table 3.20. Processing conditions for BAK [99]

	BAK 2195	BAK 1095
Melting point	175°C	125°C
Mass temperature	180–200°C	140–220°C
Mould temperature	50°C	30–40°C
Deforming	Good	Reasonable
Fogging	No	No
Corrosion	No	No
Drying conditions	2 h at 70°C	2 h at 90°C

Applications

Potential applications for BAK 1095 resin include uses in the horticulture, agriculture and food sectors. Specific examples are: biowaste bags, agricultural films, plant pots, plant clips, cemetery decoration, one-way dishes [99].

3.2.5 Poly(vinyl alcohol) – PVA

Properties

Poly(vinyl alcohol) (PVA) is a water soluble polymer based on petroleum resources with interesting properties such as good transparency, lustre, antielectrostatic properties, chemical resistance and toughness [100]. It has also good gas barrier properties and good printability. The final properties of PVA depend on the properties of its parent polymer, i.e. poly(vinyl acetate), its polymerization conditions and degree of hydrolysis. Basic properties of PVA and PVA-based systems also depend upon the degree of polymerization, distribution of hydroxyl groups, stereoregularity and crystallinity [101]. For example, the degrees of hydrolysis and polymerization affect the solubility of PVA in water [102]. PVA grades with high degrees of hydrolysis have low solubility in water. The presence of acetate groups affects the ability of PVA to crystallize upon heat treatment [102]. PVA grades containing high degrees of hydrolysis are more difficult to crystallize.

Commercial PVA grades are available with various degrees of hydrolysis and polymerization.

Processing

Two technologies are used for PVA film production – casting from viscous water solution or blown extrusion from melt. Plastic items based on PVA film are mainly obtained using casting

techniques [103]. However, due to interest in biodegradable PVA-based film, melt processing technology has been developed. The main difficulty in PVA thermal extrusion processing
is the close proximity of its melting point and decomposition temperature [103]. The thermal
degradation of PVA usually starts at about 150°C or above, depending upon the PVA grade
(degree of hydrolysis and pH). In order to improve the thermal stability and processing properties of PVA the use of plasticizers is required. Various plasticizers such as water, glycerol,
ethylene glycol and its dimer and trimer, amine alcohols, and polyvalent hydroxyl compounds
have been applied.

Applications
Poly(vinyl alcohol) is largely used as fibre, film, in the paper industry, in textile sizing, as a
modifier of thermosetting resins, in plywood manufacture, as pressure sensitive adhesives, as
an emulsifier [100, 101]. Mainly it is used as a sizing agent or stabilizer of dispersion systems.
In particular the four major segments of PVA consumption comprise: warp sizing, paper coating, coatings, and films [101].

PVA applications include textile sizing agents, paper processing agents, emulsification dis-
persants, films and general industrial use, in particular:

- Textile sizing and finishing
- Laminating adhesives
- Size in paper and paperboard manufacture
- Water-soluble films for packaging and release applications
- Protective colloid in emulsion polymerization processes
- Photosensitive coatings
- Binders for building products such as ceramics, ceiling tiles, floor coatings and paper board
- Binders for pigmented paper coatings, ceramic materials and non-woven fabrics

3.3. BLENDS

In order to obtain the compostable polymer materials with the best compromise between
mechanical and processing properties and cost, as well as compostability, various blends of
biodegradable polymers have been studied. For example, blends such as PLA/PHA, PLA/
starch have improved performance with respect to degradation rate, permeability characteristics, and thermal and mechanical properties. Overall processability is thus improved and the
range of possible applications for PLA is broadened. Blends of PLA and natural fibres have
increased durability and heat resistance and resulted in a lower cost to weight ratio compared
to unblended PLA [6].

All possible systems, including blends of polymers based on renewable and petrochemical
resources have been developed. However, most attention is given to starch-based blends [39,
104–106]. Starch is one of the most inexpensive and most readily available of compostable
polymers. Renewability of starch is another advantage of starch-based blends. The major drawback of thermoplastic starch (TPS) is sensitivity to water and poor mechanical properties. TPS
is a very hydrophilic material. To overcome moisture sensitivity and changes in mechanical
properties of TPS in relation to the crystallinity and the contents of plasticizer and water, during ageing, blending TPS with other biopolymers has been commonly used [104]. Associations
between TPS and other biopolymers include aliphatic polyesters such as polycaprolactone

Table 3.21. Properties of commercially available starch-based blends [6, 106]

	Starch (>85%)/co-polyester Mater-Bi NF01U [6]	Starch/PCL Mater-Bi ZF03U/A [6]	Starch/cellulose acetate Mater-Bi Y1010U [106]	Starch/cellulose acetate Bioplast GF105/30 [6]	Modified starch Cornpol [6]
Physical properties					
Melt flow rate (g/10 min)	2–8		10–15	5–9	5–6
Density (g/cm^3)	1.3	1.23	1.35	1.21	1.2
Transparency (%)					
Mechanical properties					
Tensile strength at yield (MPa)	25	31	25–30	44, 38	30
Elongation at yield (%)	600	900	2–6	400, 500	600–900
Tensile modulus (MPa)	120	180	2100–2500		10–30
Thermal properties					
HDT (°C)					85–105
Vicat softening point (°C)				65	105–125
GTT (°C)					
Melting point (°C)	110	64			

Samples aged two weeks at 23°C and 50% RH.

(PCL), polylactic acid (PLA), polyhydroxybutyrate-*co*-valerate (PHBV), and polyesteramide. Some starch-based blends have been commercialized such as Mater-Bi (Novamont) or Bioplast (Biotec).

The properties of commercially available starch-based blends are summarized in Table 3.21.

Different compositions of wheat thermoplastic starch (TPS) and polycaprolactone (PCL) were melt blended by extrusion and injected [104]. It was noticed that the addition of PCL to the TPS matrix allowed the weakness of pure TPS to be overcome: low resilience, high moisture sensitivity and high shrinkage, even at low PCL concentrations, e.g. 10 wt%. However, a fairly low compatibility between both polymeric systems was reported. For PCL-based blends, mechanical properties depend both on plasticization level and PCL content (Table 3.22).

PCL, due to its low melting point (~65°C), is difficult to process by conventional techniques for thermoplastic materials. Blending of starch with PCL improves its processability and furthermore promotes its biodegradation. Poly(ε-caprolactone)/plasticized starch blends varying in starch content were processed by conventional extrusion, injection moulding, and film blowing techniques [107]. Blending plasticized starch with PCL, increased the modulus and decreased the other mechanical properties (i.e. strength and elongation at yield and break) of both injection moulded specimens and films.

The processability, mechanical and thermal properties, and biodegradability of poly(butylene succinate adipate) (PBSA)/starch films containing up to 30 wt% corn starch were studied [108]. Increasing the starch content led to an increase in modulus and decrease in tensile strength, elongation to break and toughness.

Table 3.22. Mechanical properties of TPS/PCL blends [104]

PCL, wt%	TPS formula (components in wt%)	Modulus, MPa	Maximum tensile strength, MPa	Elongation at break, %	Impact strength, kJ/m^2
100	–	190	14.2	>550	No break
0	Starch 74/glycerol 10/water 16	997	21.4	3.8	0.63
25	Starch 74/glycerol 10/water 16	747	10.5	2.0	1.57
40	Starch 74/glycerol 10/water 16	585	9.0	2.4	2.99
0	Starch 70/glycerol 18/water 12	52	3.3	126.0	No break
25	Starch 70/glycerol 18/water 12	93	5.9	62.6	No break
0	Starch 67/glycerol 24/water 9	26	2.6	110.0	No break
25	Starch 67/glycerol 24/water 9	80	5.3	42.2	No break
0	Starch 65/glycerol 35	2	0.61	90.7	No break
10	Starch 65/glycerol 35	8	1.05	61.9	No break
25	Starch 65/glycerol 35	36	2.87	43.1	No break
40	Starch 65/glycerol 35	71	5.19	50.4	No break

Mechanical properties of TPS blended with poly(hydroxy butyrate) (PHB) confer higher performance than those of pristine TPS [109]. In particular, a significant increase in tensile strength and tear strength is observed for TPS (potato starch) blended with PHB at low gelatinization degree. For example, for TPS blended with 7% PHB tear strength reaches 44.1 kJ/m^2, a 12-fold increase compared with unfilled TPS at 25% glycerol content.

The properties of blends of starch and aliphatic biodegradable polyesters, including poly(ε-caprolactone), poly(butylene succinate) and a butanediol-adipate-terephthalate copolymer were studied [110]. To improve the compatibility between the starch and the synthetic polyester, a compatibilizer containing an anhydride functional group incorporated into the polyester backbone was used. The addition of a small amount of compatibilizer increased the strength significantly over the uncompatibilized blend. For the compatibilized blend, the tensile strength was invariant with starch content when compared to the original polyester, while it decreased with increase in starch content for the uncompatibilized blend.

The interfacial interaction between poly(lactic acid) (PLA) and starch was improved and mechanical properties of PLA blends with starch were enhanced by an addition of methylene diisocyanate (MDI) [111–112].

The blending of thermoplastic starch with other biodegradable polyesters such as polyesteramide could be an interesting way to produce new biodegradable starch-based materials [113]. A range of blends was studied with glycerol (plasticizer)/starch contents ratios varying from 0.14 to 0.54 [113]. BAK polyesteramide concentrations were up to 40 wt%, TPS remaining as the major phase in the blend. It was reported that the addition of BAK to TPS matrix allowed the weaknesses of pure thermoplastic starch to be overcome: low mechanical properties, high moisture sensitivity, and high shrinkage in injection, even at 10 wt% BAK. Tensile yield properties of polyesteramide (PEA) blended with granular corn starch or potato starch over a range of strain rates were investigated [114]. Yield stress increased relative to the unfilled PEA with starch volume fraction and stress rate when corn starch was the filler. When potato starch was used, the yield stress decreased with starch volume fraction at low strain rates, and increased at high strain rates.

Applications
Commercially available starch-based blends (Novamont Mater-Bi) depending on the grade are used in the following areas [106]:

Mater-Bi Z class
mainly for films and sheets
Technology: film blowing (ZF03U/A)
Use: bags, nets, paper lamination, mulch films, twines, wrapping film

Mater-Bi Y class
for rigid and dimensionally stable injection moulded items
Technology: injection moulding
Use: cutlery, boxes, flowers pots, seedling plant trays, golf tees, vending cups, pens

Mater-Bi V class
for rigid and expanded items
Technology: foaming
Use: loose fillers and packaging foams as a replacement for polystyrene
Technology: injection moulding
Use: soluble cotton swabs, soluble items

In general, the main applications of starch-based materials include [6]:

- Packaging: leaf collection compost bags, packaging films, shopping bags, strings, straws, tableware, tapes, technical films, trays and wrap film
- Agricultural sector: mulch film, planters, planting pots, encapsulation and slow release of active agents such as agrochemicals
- Transportation: fillers in tyres
- Miscellaneous: nappy back sheets, soluble cotton swabs, soluble loose fillers, cups, cutlery, edge protectors, golf tees, mantling for candles and nets

3.4. SUMMARY

Physical and mechanical properties of the main compostable polymers are summarized in Tables 3.23 and 3.24.

Table 3.23. Summary of properties of compostable polymer materials derived from renewable resources [115]

Property	PLA	L-PLA	DL-PLA	PHB
Density, g/cm^3	1.21–1.25	1.24–1.30	1.25–1.27	1.18–1.262
Glass transition, °C	45–60	55–65	50–60	5–15
Melting temperature, °C	150–162	170–200	amorphous	168–182
Tensile strength, MPa	21–60	15.5–150	27.6–50	40
Tensile modulus, GPa	0.35–3.5	2.7–4.14	1–3.45	3.5–4
Ultimate tensile strain, %	2.5–6	3–10	2–10	5–8

Table 3.24. Summary of properties of compostable polymer materials derived from petrochemical resources [116]

Property	Polycaprolactone (PCL)	Polyesteramide (PEA)	Polybutylene succinate/adipate (PBSA)	Polybutylene adipate-*co*-terphthalate (PBAT)
Density, g/cm^3	1.11	1.05	1.23	1.21
Glass transition, °C	−60	−30	−45	−30
Melting temperature, °C	60–62	125	92–94	110–115
Tensile strength, Mpa	20.7–42			
Tensile modulus, Gpa	0.21–0.44			
Ultimate tensile strain, %	300–1000			

Applications of compostable polymer materials commercialized or in the development/demonstration stage include [106]:

- Packaging: films and trays for biscuits, fruit, vegetables and meat, yoghurt cup, nets for fruit, grocery bags, rigid transparent packaging of batteries with removable printed film on back side, trays and bowls for fast food, envelopes with transparent window, paper bags for bread with transparent window
- Agriculture and horticulture: mulching films, tomato clips
- Short life consumer goods: hygiene products such as nappies, cotton swabs, stationary and pre-paid cards
- Longlife consumer goods: apparel, e.g. T-shirts, socks, blankets, mattresses, casings for Walkman, CDs (compact discs), computer keys, small components of laptop housing, spare wheel covers, automobile interiors including head liners and upholstery and possibly for trimmings

3.4.1 Major markets of compostable polymer materials
- Agricultural and fishery (mulch films, pots for transplanting, fishing lines and nets)
- Civil engineering and construction (sand bags, flora sheets, curing sheets)
- Leisure goods (golf tees, marine sports and mountain climbing)
- Food packaging (trays for perishable food, fast-food containers)
- Packaging (kitchen garbage, composting bags, bin liner bags, shopping bags)
- Textile goods (clothes, mats)
- Daily use (pen cases, disposal shavers)
- Electronic (electronic equipment cases)
- Automotive industry (car parts)

The newest application trends are described in Chapter 9.

REFERENCES

[1]   Vink E.T.H., Rábago K.R., Glassner D.A., Springs B., O'Connor R.P., Kolstad J., Gruber P.R.: The sustainability of NatureWorks polylactide polymers and Ingeo polylactide fibers: an update of the future. Initiated by the 1st International Conference on Bio-based Polymers (ICBP 2003), 12–14 November 2003, Saitama, Japan, *Macromol. Biosci.* 4 (2004) 551.

[2]   Jamshidi K., Hyon S.-K, Ikada Y.: Thermal characterization of polylactides, *Polymer* 29 (1988) 2229.

[3]   Martin O., Avérous L.: Poly(lactic acid): plasticization and properties of biodegradable multiphase systems, *Polymer* 42 (2001) 6209.

[4]   Lunt J.: Large-scale production, properties and commercial applications of polylactic acid polymers, *Polym. Deg. Stab.* 59 (1998) 145.

[5]   Södergård A., Stolt M.: Properties of lactic acid based polymers and their correlation with composition, *Prog. Polym. Sci.* 27 (2002) 1123.

[6]   Techno-economic feasibility of large-scale production of bio-based polymers in Europe (PRO-BIP), Final Report, Utrecht/Karlsruhe, October 2004.

[7]   Drumright R.E., Gruber P.R., Henton D.E.: Polylactic acid technology, *Adv. Mater.* 12 (2000) 1841.

[8]   Source: NatureWorks; www.natureworksllc.com

[9]   Source: Hycail; www.hycail.com

[10]   Nakagawa T., Nakiri T., Hosoya R., Tajitsu Y.: Electrical properties of biodegradable polylactic acid films, *Proceedings of the 7th International Conference on Properties and Applications of Dielectric Materials*, 1–5 June 2003, Nagoya, Japan.

[11]   Additives for Polymers, March 2000, 8.

[12]   Khanna S., Srivastava A.K.: Recent advances in microbial polyhydroxyalkanoates, *Process Biochemistry* 40 (2005) 607.

[13]   Braunegg G., Lefebre G., Genser K.F.: Polyhydroxyalkanoates, biopolyesters from renenewable resources: physiological and engineering aspects, *J. Biotechnology* 65 (1998) 127.

[14]   Lee S.Y.: Bacterial polyhydroxyalkanoates, *Biotechnol. Bioeng.* 49 (1996) 1.

[15]   Padermshoke A., Katsumoto Y., Sato H., Ekgasit S., Noda I., Ozaki Y.: Melting behaviour of poly(3-hydroxybutyrate) investigated by two-dimensional infrared correlation spectroscopy, *Spectrochimica Acta Part A* 61 (2005) 541.

[16]   Kunioka M., Tamaki A., Doi Y.: Crystalline and thermal properties of bacterial copolyesters: Poly(3-hydroxybutyrate-*co*-3-hydroxyvalerate) and poly(3-hydroxybutyrate-*co*-4-hydroxybutyrate), *Macromolecules* 22 (1989) 694.

[17]   Suriyamongkol P., Weselake R., Narine S., Moloney M., Shah S.: Biotechnological approaches for the production of polyhydroxyalkanoates in microorganisms and plants – a review, *Biotechnol. Adv.* 25 (2007) 148.

[18]   Narasimhan K., Green P.: Nodax™ Biopolymer. Creating the low-cost supply for PHA-polyhydroxyalkanoates, ACS – 1 April 2004.

[19]   Rabetafika H.N., Paquot M., Janssens L., Castiaux A., Dubois Ph.: Development durable et resources renouvelables. Rapport final, PADD II, 2006, Bruxelles.

[20]   Source: www.nodax.com.

[21]   Reddy C.S.K. Ghai R., Rashmi, Kalia V.C.: Polyhydroxyalkanoates: an overview, *Bioresource Technology* 87 (2003) 137.

[22]   Poutanen K., Forsell P.: Modification of starch properties with plasticizers, *TRIP* 4 (1996) 128.

[23]   Avérous L., Fringant C.: Association between plasticized starch and polyesters: processing and performances of injected biodegradable systems, *Polym. Eng. Sci.* 41 (2001) 727.

[24]   Van Soest J.J.G., Benes K., de Wit D.: The influence of starch molecular mass on the properties of extruded thermoplastic starch, *Polymer* 37 (1996) 3543.

[25]   Van Soest J.J.G., Hulleman S.H.D., de Wit D., Vliegenthart J.F.G.: Changes in the mechanical properties of thermoplastic starch in relation with changes in B-type crystallinity, *Carbohydr. Polym.* 29 (1996) 225.

[26]   Van Soest J.J.G., de Wit D., Vliegenthart J.F.G.: Mechanical properties of thermoplastic maize starch, *J. Appl. Polym. Sci.* 61 (1996) 1927.

[27]   Forsell P.M., Mikkilä J.M., Moates G.K., Parker R.: Phase and glass transition behaviour of concentrated barley starch-glycerol-water mixtures, a model for thermoplastic starch, *Carbohydr. Polym.* 34 (1997) 275.

[28]   Rodriguez-Gonzalez F.J., Ramsay B.A., Favis B.D.: Rheological and thermal properties of thermo-plastic starch with high glycerol content, *Carbohydr. Polym.* 58 (2004) 139.

[29]   De Graaf R.A., Karman A.P., Janssen L.P.B.M.: Material properties and glass transition tempera-tures of different thermoplastic starches after extrusion processing, *Starch* 55 (2003) 80.

[30]   Huang M., Yu J., Ma X.: Etanoloamine as a novel plasticizer for thermoplastic starch, *Carbohydr. Polym.* 90 (2005) 501.

[31]   Da Roz A.L., Carvalho A.J.F., Gandini A., Curvelo A.A.S.: The effect of plasticizers on thermoplastic starch compositions obtained by melt processing, *Carbohydr. Polym.* 63 (2006) 417.

[32]   Thunwall M., Boldizar A., Rigdahl M.: Compression molding and tensile properties of thermo-plastic potato starch materials, *Biomacromolecules* 7 (2006) 981.

[33]   Ma X., Yu J.: The plasticizers containing amide groups for thermoplastic starch, *Carbohydr. Polym.* 57 (2004) 197.

[34]   Ma X., Yu J., Ma Y.: Urea and formamide as a mixed plasticizer for thermoplastic wheat flour, *Carbohydr. Polym.* 60 (2005) 111. Yang J.-H., Yu J.-G., Ma X.: Preparation and properties of ety-lenebisformamide, *Carbohydr. Polym.* 63 (2006) 218.

[35]   Shogren R.L., Swanson C.L., Thompson A.R.: Extrudates of corn starch with urea and glycols; structure/mechanical property relations, *Starch* 44 (1992) 335.

[36]   Ma X., Yu J., Wan J.: Urea and etanoloamine as a mixed plasticizer for thermoplastic starch, *Carbohydr. Polym.* 64 (2006) 267.

[37]   Yang J., Yu J., Ma X.: Study on the properties of ethylenebisformamide and sorbitol plasticized corn starch, *Carbohydr. Polym.* 66 (2006) 110.

[38]   Hulleman S.H.D., Janssen F.H.P., Feil H.: The role of water during plasticization of native starches, *Polymer* 39 (1998) 2043.

[39]   Avérous L.: Biodegradable multiphase systems based on thermoplastic starch: a review, *Polymer Rev.* 44 (2004) 231.

[40]   Van Soest J.J.G., Kortleve P.M.: The influence of maltodextrins on the structure and properties of compression-molded starch plastic sheets, *J. Appl. Polym. Sci.* 74 (1999) 2207.

[41]   Van Soest J.J.G., Vliegenthart J.F.G.: Crystallinity in starch plastics: consequences for material properties, *TIBTECH* 15 (1997) 208.

[42]   Van Soest J.J.G.: The development of fully biodegradable starch plastics: process-structure-property relationships, *Agro-Food Industry Hi-Tech* (1997) 17.

[43]   Stepto R.F.T.: Understanding the processing of thermoplastic starch, *Macromol. Symp.* 245–246 (2006) 571.

[44]   Klemm D., Heublein B., Fink H.-P., Bohn A.: Cellulose: fascinating biopolymer and sustainable raw material, *Angew. Chem. Int. Ed.* 44 (2005) 3358.

[45]   Mohanty A.K., Wibowo A., Misra M., Drzal L.T.: Development of renewable resource-based cel-lulose acetate bioplastic: effect of process engineering on the performance of cellulosic plastics, *Polym. Eng. Sci.* 43 (2003) 1151.

[46]   Joly N., Granet R., Branland P., Verneuil B., Krausz P.: New method for acylation of pure and saw-dust-extracted cellulose by fatty acid derivatives – thermal and mechanical analysis of cellulose-based plastic films, *J. Appl. Polym.Sci.* 97 (2005) 1266.

[47]   Edgar K.J., Buchanan C.M., Debenham J.S., Rundquist P.A., Seiler B.D., Shelton M.C., Tindall D.: Advances in cellulose ester performance and application, *Prog. Polym. Sci.* 26 (2001) 1605.

[48]   www.ides.com

[49]   Srinivasa P.C., Tharanathan R.N.: Chitin/chitosan-safe, ecofriendly packaging materials with mul-tiple potential uses, *Food Rev. Int.* 23 (2007) 53.

[50]   Di Martino A., Sittinger M., Risbud M.V.: Chitosan: A versatile biopolymer for orthopaedic tissue engineering, *Biomaterials* 26 (2005) 5983.

[51]   Kumar M.N.V.: A review of chitin and chitosan applications, *Reactive & Functional Polymers* 46 (2000) 1.

[52] Srinivasa P.C., Ramesh M.N., Tharanathan R.N.: Effect of plasticizers and fatty acids on mechanical and permeability characteristics of chitosan films, *Food Hydrocoll.* 21 (2007) 1113.

[53] Rinaudo M.: Chitin and chitosan: properties and applications, *Prog. Polym. Sci.* 31 (2006) 603.

[54] Kurita K.: Controlled functionalisation of the polysaccharide chitin, *Prog. Polym. Sci.* 26 (2001) 1921.

[55] Swain S.N., Biswal S.M., Nanda P.K., Nayak P.L.: Biodegradable soy-based plastics: opportunities and challenges, *J. Polym. Environm.* 12 (2004) 35.

[56] Zheng H., Tan Z., Zhan Y., Huang J.: Morphology and properties of soy protein plastics modified with chitin, *J. Appl. Polym. Sci.* 90 (2003) 3676.

[57] Mo X., Sun X.: Effects of storage time on properties of soy protein-based plastics, *J. Polym. Environm* 11 (2003) 15.

[58] Pommet M., Redl A., Guilbert S., Morel M.: Intrinsic influence of various plasticizers on functional properties and reactivity of wheat gluten thermoplastic materials, *J. Cereal Sci.* 42 (2005) 81.

[59] Micard V., Guilbert S.: Thermal behaviour of native and hydrophobized wheat gluten, gliadin and glutenin-rich fractions by modulated DSC, *Int. J. Biol. Macromol.* 27 (2000) 229.

[60] Di Gioia L., Cuq B., Guilbert S.: Thermal properties of corn gluten meal and its proteic components, *Int. J. Biol. Macromol.* 24 (1999) 341.

[61] Zhang J., Mungara P., Jane J.: Mechanical and thermal properties of extruded soy protein sheets, *Polymer* 42 (2001) 2569.

[62] Vaz C.M., van Doeveren P.F.N.M., Yilmaz G., de Graaf L.A., Reis R.L., Cunha A.M.: Processing and characterization of biodegradable soy plastics: effects of crosslinking with glyoxal and thermal treatment, *J. Appl. Polym. Sci.* 97 (2005) 604.

[63] Zhang X., Do M.D., Hoobin P., Burgar I.: The phase composition and molecular motions of plasticized gluten-based biodegradable polymer materials studied by solid-state NMR spectroscopy, *Polymer* 47 (2006) 5888.

[64] Mangavel C., Barbot J., Guéguen J., Popineau Y.: Molecular determinants of the influence of hydrophilic plasticizers on the mechanical properties of cast wheat gluten films, *J. Agric. Food Chem.* 51 (2003) 1447.

[65] Larré C., Desserme C., Barbot J., Gueguen J.: Properties of deamidated gluten films enzymatically cross-linked, *J. Agric. Food Chem.* 48 (2000) 5444.

[66] Orliac O., Rouilly A., Silvestre F., Rigal L.: Effects of additives on the mechanical properties, hydrophobicity and water uptake of thermo-moulded films produced from sunflower protein isolate, *Polymer* 43 (2002) 5417.

[67] Orliac O., Rouilly A., Silvestre F., Rigal L.: Effects of various plasticizers on the mechanical properties, water resistance and aging of thermo-moulded films made from sunflower protein isolate, *Polymer* 18 (2003) 91.

[68] Grevellec J., Marquié C., Ferry L., Crespy A., Vialettes V.: Processability of cottonseed proteins into biodegradable materials, *Biomacromolecules* 2 (2001) 1104.

[69] Mo X., Sun X.S., Wang Y.: Effects of molding temperature and pressure on properties of soy protein polymers, *J. Appl. Polym. Sci.* 13 (1999) 2595.

[70] Hochstetter A., Tajla R., Helén H.J., Hyvönen L., Jouppila K.: Properties of gluten-based sheet produced by twin-screw extruder, *LWT* 39 (2006) 893.

[71] Irissin-Mangata J., Bauduin G., Boutevin B., Gontard N.: New plasticizers for wheat gluten films, *Eur. Polymer J.* 37 (2001) 1533.

[72] Kumar R., Choudhary V., Mishra S., Varma I.K., Mattiason B.: Adhesives and plastics based on soy protein products, *Ind. Crops Prod.* 16 (2002) 155.

[73] Day L., Augustin M.A., Batey I.L., Wrigley C.W.: Wheat-gluten uses and industry needs, *Trend. Food Sci.* 17 (2006) 82.

[74] IENICA (Interactive European Network for Industrial Crops and their Applications), Summary Report for European Union. Protein Crops. European Report 1996–2000 (available at www.ienica.net).

[75]   Fujimaki T.: Processability and properties of aliphatic polyesters, "BIONOLLE", synthesised by polycondensation reaction, Polym. Deg. Stab. 59 (1998) 209.

[76]   Uesaka T., Nakane K., Maeda S., Ogihara T., Ogata N.: Structure and physical properties of poly(butylene) succinates/cellulose acetate blends, *Polymer* 41 (2000) 8449.

[77]   Sinha Ray S., Bousmina M.: Biodegradable polymers and their layered silicate nanocomposites: in greening the 21st century materials world, *Prog. Mater. Sci.* 50 (2005) 962.

[78]   Jin H., Park J., Park K., Kim M., Yoon J.: Properties of aliphatic polyesters with n-paraffinic side branches, *J. Appl. Polym. Sci.* 77 (2000) 547.

[79]   Tserki V., Matzinos P., Pavlidou E., Vachliotis D., Panayiotou C.: Biodegradable aliphatic polyesters. Part I. Properties and biodegradation of poly(butylene succinate-*co*-butylene adipate), *Polym. Deg. Stab.* 91 (2006) 367.

[80]   Tserki V., Matzinos P., Pavlidou E., Vachliotis D., Panayiotou C.: Biodegradable aliphatic polyesters. Part II. Synthesis and characterization of chain extended poly(butylene succinate-*co*-butylene adipate), *Polym. Deg. Stab.* 91 (2006) 377.

[81]   Papageorgiou G.Z., Bikiaris D.N: Crystallization and melting behaviour of three poly(alkylene succinates), *Polymer* 46 (2005) 12081.

[82]   Kalakkunnath S., Kalika D.S.: Dynamic mechanical and dielectric relaxation characteristics of poly(trimethylene terephthalate), *Polymer* 47 (2006) 7085.

[83]   Pyda M., Boller A., Grebowicz J., Chuah H., Lebedev B.V., Wunderlich B.: Heat capacity of poly(trimethylene terephthalate), *J. Polym. Sci., Part B Polym. Phys.* 36 (1998) 2499.

[84]   Gan Z., Kuwabara K., Yamamoto M., Abe H., Doi Y.: Solid-state and thermal properties of aliphatic-aromatic poly(butylene adipate-*co*-butylene terephthalate) copolyesters., *Polym. Deg. Stab.* 83 (2004) 289.

[85]   Shi X.Q., Ito H., Kikutani T.: Characterization on mixed-crystal structure and properties of poly(butylene adipate-*co*-terephthalate) biodegradable fibers, *Polymer* 46 (2005) 11442.

[86]   Mecking S.: Nature or petrochemistry? Biologically degradable materials, *Angew. Chem. Int. Ed.* 43 (2004) 1078.

[87]   Wu G., Li H., Wu Y., Cuculo J.A.: Structure and property studies of poly(trimethylene terephthalate) high-speed melt spun fibers, *Polymer* 43 (2002) 4915.

[88]   Shafee E.El.: Effect of aging on the mechanical properties of cold-crystallized poly(trimethylene terephthalate), *Polymer* 44 (2003) 3727.

[89]   Funabashi M., Ninomiya F., Kunioka M.: Biodegradation of polycaprolactone powders proposed as reference test materials for international standard of biodegradation evaluation method, *J. Polym. Environ.* 15 (2007) 7.

[90]   Koening M.F., Huang S.J.: Biodegradable blends and composites of polycaprolactone and starch derivatives, *Polymer* 36 (1995) 1877.

[91]   Matzinos P., Tserki V., Kontoyiannis C., Panayiotou C.: Processing and characterization of starch/polycaprolactone products, *Polym. Deg. Stab.* 77 (2002) 17.

[92]   Source: Solvay; www.solvay.com.

[93]   Source: Union Carbide, www.unioncarbide.com.

[94]   Edlund U., Albertsson A.-Ch.: Degradable polymer microspheres for controlled drug delivery in: Degradable aliphatic polyesters, *Adv. Polym. Sci. Vol.* 157, Springer 2002.

[95]   Botines E., Rodríguez-Galán A., Puiggalí J.: Poly(ester amide)s derived from 1,4-butanediol, adipic acid and 1,6-aminohexanoic acid: characterization and degradation studies, *Polymer* 43 (2002) 6073.

[96]   Ferré T., Franco L., Rodríguez-Galán A., Puiggalí J.: Poly(ester amide)s derived from 1,4-butanediol, adipic acid and 1,6-aminohexanoic acid. Part II: composition changes and fillers, *Polymer* 44 (2003) 6139.

[97]   Lozano M., Franco L., Rodríguez-Galán A., Puiggalí J.: Poly(ester amide)s derived from 1,4-butanediol, adipic acid and 1,6-aminohexanoic acid. Part III: substitution of adipic acid units by terephthalic acid units, *Polym. Deg. Stab.* 85 (2004) 595.

[98] Lips P.A.M., Broos R., Van Heeringen M.J.M., Dijkstra P.J., Feijen J.: Synthesis and characterization of poly(ester amide)s containing crystallizable amide segments, *Polymer* 46 (2005) 7823.

[99] Grigat E, Koch R., Timmermann R: BAK 1095 and BAK 2195: completely biodegradable synthetic thermoplastics , *Polym. Deg. Stab.* 59 (1998) 223.

[100] Gohil J.M., Bhattacharya A., Ray P.: Studies on the cross-linking of poly(vinyl alcohol), *J. Polym. Res.* 13 (2006) 161.

[101] Chiellini E., Corti A., Solaro R.: Biodegradation of poly(vinyl alcohol) based blown films under different environmental conditions, *Polym. Deg. Stab.* 64 (1999) 305.

[102] Hassan Ch.M., Peppas N.A.: Structure and applications of poly(vinyl alcohol) hydrogels produced by conventional crosslinking or by freezing/thawing methods, Adv. Polym. Sci. 153 (2000) 37.

[103] Chiellini E., Corti A., D'Antone S., Solaro R.: Biodegradation of poly(vinyl alcohol) based materials, *Prog. Polym. Sci.* 28 (2003) 963.

[104] Avérous L., Moro L., Dole P., Fringant C.: Properties of thermoplastic blends: starch-polycaprolactone, *Polymer* 41 (2000) 4157.

[105] Wang X.-L., Yang K-K., Wang Y.-Z.: Properties of starch blends with biodegradable polymers, *J. Macromol. Sci. Part C – Polym. Rev.* C43 (2003) 385.

[106] Bastioli C.: Properties and applications of Mater-Bi starch-based materials, *Polym. Deg. Stab.* 59 (1998) 263.

[107] Matzinos P., Tserki V., Kontoyiannis A., Panayiotou C.: Processing and characterization of starch/polycaprolactone products, *Polym. Deg. Stab.* 77 (2002) 17.

[108] Ratto J.A. Stenhouse P.J., Auerbach M., Mitchell J., Farrell: Processing, performance and biodegradability of a thermoplastic aliphatic polyester/starch system, *Polymer* 40 (1999) 6777.

[109] Lai S.-M., Don T.-M., Huang Y.-C.: Preparation and properties of biodegradable thermoplastic starch/poly(hydroxy butyrate) blends, *J. Appl. Polym. Sci.* 100 (2006) 2371.

[110] Mani R., Bhattacharya M.: Properties of injection moulded blends of starch and modified biodegradable polyesters, *Eur. Polym. J.* 37 (2001) 515.

[111] Wang H., Sun X., Seib P.: Strengthening blends of poly(lactic acid) and starch with methylene diisocyanate, *J. Appl. Polym. Sci.* 82 (2001) 1761.

[112] Wang H., Sun X., Seib P.: Properties of poly(lactic acid) blends with various starches as affected by physical aging, *J. Appl. Polym. Sci.* 90 (2003) 3683.

[113] Avérous L., Fauconnier N., Moro L., Fringant C.: Blends of thermoplastic starch and polyesteramide: processing and properties, *J. Appl. Polym. Sci.* 76 (2000) 1177.

[114] Willett J.L., Felker F.C.: Tensile yield properties of starch-filled poly(ester amide) materials, *Polymer* 46 (2005) 3035.

[115] Van de Velde K., Kiekens P.: Biopolymers: overview of several properties and consequences on their applications, *Polymer Testing* 21 (2002) 433.

[116] Avérous L., Fringant C.: Association between plasticized starch and polyesters: processing and performances of injected biodegradable systems, *Polym. Eng. Sci.* 41(2001) 727.

Chapter 4
Thermal and thermooxidative degradation

Chapter 4
Thermal and thermooxidative degradation

When the degradability of polymeric materials in a composting environment is considered, the primary degradation mechanism is thought to be biodegradation [1]. Essentially in this process microorganisms such as bacteria and fungi degrade the material producing CO_2 and other natural products. However, because of the elevated temperatures associated with the commercial composting process (i.e. 60–65°C), coupled with the presence of moisture and oxygen, other chemical degradation processes can possibly occur. Thus, it is important to investigate the thermal behaviour of compostable polymer materials, including thermooxidative degradation processes, occurring in an oxidizing environment. From the perspective of stabilizing the polymer during processing and product use, thermal properties such as thermal stability of the material are of great importance. In order to gain knowledge about the thermal behaviour of polymeric materials in various environments it is also noteworthy to study their thermal properties in an inert atmosphere, e.g. a thermal degradation process.

4.1. BIODEGRADABLE POLYMERS FROM RENEWABLE RESOURCES

4.1.1 Polylactide – PLA
The thermal stability of aliphatic polyesters is in general limited [2]. Polylactides were found to be highly sensitive to heat, especially at temperatures higher than190°C [3]. There may be several reasons for its poor thermal stability: (1) hydrolysis by trace amounts of water catalysed by hydrolysed monomer (lactic acid); (2) zipper-like depolymerization, catalysed by residual polymerization catalyst; (3) oxidative, random main-chain scission; (4) intermolecular transesterification to monomer and oligomeric esters; and (5) intramolecular transesterfication resulting in formation of the monomer and oligomeric lactides of low molecular weight.

The dominant reaction pathway is an intramolecular transesterification (T_{max} = 360°C) giving rise to formation of cyclic oligomers [4]. In addition, acrylic acid from *cis*-elimination as well as carbon oxides and acetalaldehyde from fragmentation reactions were detected.

It was suggested by Aoyaggi *et al.* [5] in accordance with previous works [6] that the thermal degradation behaviour of PLA was very complex because various reactions occurred concurrently. Py-GC/MS analysis indicated acetalaldehyde, lactide monomer and oligomers of lactide as products of PLA pyrolysis.

The parameters that have been reported to influence PLLA thermal stability include moisture, hydrolysed monomers and oligomers, molecular weight and residual metals [3]. Cam *et al.* [6] reported that the metal residues (tin, zinc, aluminium and iron salts) as additives decreased thermal stability of PLLA in the order of Sn < Zn < Al < FE. PLA is thermally unstable and exhibits a rapid loss of molecular weight as the result of thermal treatment at processing temperatures [7]. The influence of processing parameters, namely process temperature, residence time, and the inherent moisture content on the degradation of poly(L-lactide) (PLLA) has been investigated [7]. PLLA polymer was processed by melt extrusion in a double screw extruder at 210°C and 240°C. It was demonstrated that the temperature in the extruder must be kept at a low level to minimize degradation of the polymer during processing. At the lowest processing temperature used, 210°C, the loss in number-average molecular weight (M_n)

was less dependent on the residence time in the melt compared to when processed at a temperature of 240°C. The presence of moisture in the material affected the loss in M_n to a great extent when processing was done at 210°C.

Poly(D-lactide) (PDLA) and poly(D-lactide) and their equimolar enantiometric blend (PLLA/PDLA) films were prepared and the effects of enantiometric polymer blending on the thermal stability and degradation of the films were investigated isothermally and non-isothermally under nitrogen gas using thermogravimetry (TG) [8]. At the temperature below 260°C the enantiometric polymer blending was found to successfully enhance the thermal stability of the PLLA/PDLA film compared with those of the pure PLLA and PDLA films. It was suggested that the enhancement can be ascribed to the peculiar strong interactions between PLLA and PDLA chains in the film even when they are in the melt, which decreases the mobility of the chains and thereby retards the thermal degradation of the film.

It was observed that notable factors influencing the properties and processing of racemic copolymers of lactic acid were the concentration of residual monomer in the polymer, the processing temperature history, and the extent of molecular weight degradation during processing [9]. Thermal degradation considerations established a maximum processing temperature of 200°C. This upper processing temperature results in a very narrow processing window, approximately 12°C, for 100% (L)-PLA. The 90/10 (L-)/(D, L-) copolymer has a broader processing window (40°C), by virtue of its lower melting point (150°C). It was confirmed that end capping reduced the rate of degradation. The end-capped polymers degraded at less than half the rate of the uncapped polymer. End capping not only stabilizes the polymer, but inhibits colour formation during melt processing. Colour formation can also be prevented by plasticizing PLA with at least 15% lactide monomer [9].

A mathematical model to describe the molecular weight and polydispersity index in poly(L-lactide) (PLLA) thermal degradation has been developed [10]. Based on the random scission mechanism, effects of temperature and time on the molecular weight and polydispersity index are included in this model.

4.1.2 Polyhydroxyalkanoates – PHA

Poly(β-hydroxybutyrate) PHB has a high melting point (180°C) and forms highly crystalline solids which crystallize slowly and form large spherulitic structures that impart poor mechanical properties in moulded plastics and films. Because of its high melting point, PHB is also susceptible to thermal degradation during melt processing by ester pyrolysis of the aliphatic secondary esters of the repeating units [11].

PHB has low resistance to thermal degradation; it easily decomposes near its melting temperature, i.e. 180°C. P((R)-3HB) is thermally instable at temperatures above 160°C [12, 13]. During early stages of the thermal degradation at temperatures above 160°C, the degradation occurs exclusively via random chain scission (*cis*-elimination) of the ester group, which has a six-member ring ester intermediate, to form olefinic and carboxylic acid groups [13–16]. As a result, a drastic reduction in molecular weight occurs during processing at temperatures above the melting temperature.

Thermal degradation of PHB has been suggested to occur almost exclusively by a non-radical random chain scission reaction (*cis*-elimination) according to scheme given in Fig. 4.1. [5, 13–16]. The volatile products of PHB were analysed by Py-GC/MS at 280°C [5]. Crotonic acid and its dimers and trimers were found. These observations were in accordance with the proposed thermal degradation mechanism of PHB which almost exclusively involves a random chain scission (*cis*-elimination) reaction of the ester groups to form crotonic acids and its oligomers.

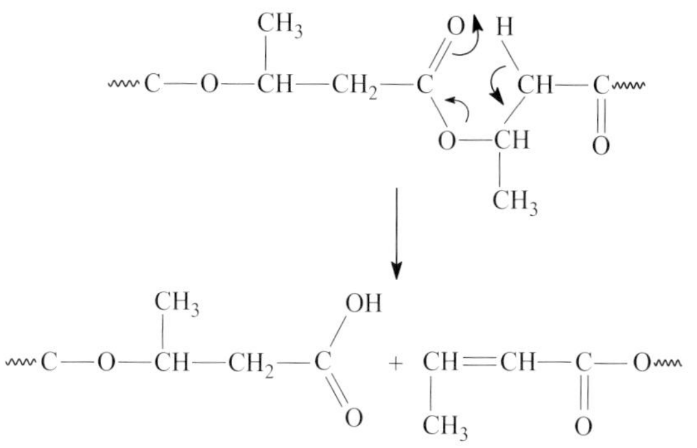

Figure 4-1 Schematic description of the thermal degradation of PHB [13].

PHB is so susceptible to thermal breakdown that pyrolysis yields of crotonic acid approach 90% [17]. To control this aspect of degradation during processing, suitable additives are used. A detailed study of thermal degradation of PHB identified the volatile products of degradation. When heated from 0 to 338°C under vacuum, PHB releases isocrotonic acid (0.9 wt%), crotonic acid (35.3 wt%) and the dimer (41.2 wt%), trimer (12.5 wt%) and tetramer (2.9 wt%) of PHB. When the heating is continued to 500°C, traces (4 wt%) of the degradation products of these volatiles are observed: carbon dioxide, propene, ketene, acetalaldehyde and β-butyrolactone.

Thermal degradation of two types of copolyesters: poly(3-hydroxybutyrate-*co*-3-hydroxy-valerate) (3HV = 0–71 mol%) and poly(3-hydroxybutyrate-*co*-4-hydroxybutyrate) (0–82%), were studied in the temperature range 100–200°C by monitoring the time-dependent changes in molecular weight of melt samples [18]. All copolyester samples were thermally unstable at temperatures above 170°C.

Thermal degradation of various polyhydoxyalkanoates, i.e. poly(3-hydroxybutyrate) (PHB), poly(3-hydroxybutyrate-*co*-3-hydroxyvalerate) (PHBV), and poly(3-hydroxybutyrate-*co*-3-hydroxyhexanoate) (PHB-HH) was investigated under nitrogen atmosphere by dynamic thermogravimetry [19]. The incorporation of 30 mol% 3-hydoroxyvalerate (HV) and 15 mol% 3-hydroxyhexanoate (HHx) components into the polyester increased the thermal stability.

In air, the thermooxidative degradation and stability of commercial and laboratory made polyhydroxyalkanoates such as the homopolymer poly(3-hydroxybutyrate) and the copolymer poly(3-hydroxybutyrate-*co*-3-hydroxyvalerate) was investigated by Carraso *et al.* [20]. It was indicated that the presence of hydroxyvalerate within the copolymer led to a thermally more stable material (with an increase of 14°C).

4.1.3 Thermoplastic starch – TPS

The degradation of cellulose and starch in air and nitrogen has been investigated by thermal analyses techniques [21]. Between ambient and 250°C only small or negligible mass loss was apparent. Some loss of adsorbed water occurred. Between 250°C and 500°C both cellulose and starch underwent a dramatic loss due to production of volatile components and formation of chars and tar. Below 360°C, in oxygen, the volatile components caught fire (gaseous combustion). Above 360°C the reaction of the solid carbonaceous residues with oxygen took place and resulted in glowing combustion. Under an oxidizing atmosphere, the reaction went to completion at approximately 550°C. Under an inert atmosphere, residual char remained at 750°C.

In general, dehydration and depolymerization have been considered as the two main processes associated with the degradation mechanism of polysaccharides [22]. In the case of starch, ether bonds and unsaturated structures are formed via thermal condensation between hydroxyl groups of starch chains, which eliminates water and other small molecules, and by dehydration of hydroxyl groups in the glucose ring. It was observed that water was the main product of decomposition of modified starch at a temperature below 300°C, formed by intermolecular or intramolecular condensation of starch hydroxyls [23, 24].

Rheological measurements in time sweep mode for thermoplastic wheat starch (TPS) with 40% glycerol content demonstrate that TPS has excellent thermal stability at 150°C, but becomes unstable at temperatures above 180°C [25]. However, it was found that TPS stability is maintained for short time periods at temperatures up to 200°C.

4.1.4 Other compostable polymers from renewable resources

Cellulose
In cellulose degradation, the dehydration process forms char, and volatiles such as CO_2, CO, H_2O, aldehydes, etc. due to scission of the glucose ring [26]. The depolymerization (higher temperatures) produces CO_2, CO, liquid products and char. The evolution of water during the heating of cellulose occurs both physically through desorption and chemically by elimination reactions [25]. The mechanisms of water evolution from cellulose in three distinct temperature regions (i.e. loss of absorbed water at low temperatures (<220°C), loss of chemical water at moderate-to-high temperatures (220–550°C) and loss of chemical water at high temperatures (>550°C) were discussed. It was suggested that chemical elimination of water from cellulose originated primarily from an intramolecular elimination leading to C2, C3 unsaturation or a ketone group on C2. Scheirs *et al.* [26] have reported that first decomposition processes in cellulose are mainly due to dehydration reactions which occur at temperatures between 210°C and 325°C.

The kinetics of the thermooxidative degradation of cellulose and its esters in air were studied by thermogravimetry and differential thermal analysis from ambient temperature to 650°C [27]. It was found that decomposition temperature and maximum degradation rates are higher for cellulose esters than for pure cellulose. However, activation energies for the second stage of degradation of cellulose esters were lower than that of pure cellulose.

Thermal degradation behaviour of some partially esterified long chain cellulose esters was studied by Jandura *et al.* [28]. Cellulose esters showed lower decomposition temperature than cellulose. However, the thermal stability of cellulose esters, with the exemption of cellulose oleate, improved with higher degree of substitution.

Chitosan
Chitosan, like other polysaccharides, is susceptible to a variety of degradation mechanisms, including oxidative/reductive free radical depolymerization and acid-, alkaline- and enzyme-catalysed hydrolysis [29]. Degradation of polysaccharides occurs via cleavage of the glycosidic bonds. Depolymerization of chitosan is useful in order to control properties such as viscosity, solubility and biological activity. Potential mechanisms for temperature-induced degradation of chitosan are oxidative/reductive degradation and acid catalysed degradation [29]. The thermal degradation of chitin and chitosan has been studied by using simultaneous DSC and TG techniques in nitrogen atmosphere [30]. The thermal degradation of chitin and chitosan follows a random scission pathway.

Proteins

During thermal treatment cereal proteins undergo irreversible aggregation reactions via disulphide crosslinks [31]. This reaction does not lead to a change in T_g value for plasticized gluten but results in the formation of a network structure, in a similar way to the curing of epoxy resin or the vulcanization of rubber. Activation energy for heat-induced aggregation of gluten, measured from its solubility decrease in sodium dodecyl sulphate (SDS) solvent, is about 172–183 kJ/mol. Two major reactions involved during gluten heating, protein aggregation and polypeptide degradation, were tested at two oscillatory strains (7%: linear domain; 70%: high deformation domain). Strain was found to moderately accelerate the gluten protein aggregation reaction and to greatly accelerate the gluten thermal degradation. Thermogravimetric analysis of moulded soy protein plastics in nitrogen gas showed that plastics were stable up to 300°C, indicating good thermal stability. With the presence of oxygen, the plastics decomposed at 180°C [32, 33].

The thermal stability of films formed by the soy protein isolate (SPI)–sodium dodecyl sulphate (SDS) complex was investigated by thermogravimetric analysis (TGA) and Fourier transform infrared spectroscopy (FTIR) in a nitrogen atmosphere [34]. The thermal degradation of SPI films occurred in a single process that began at 292°C and reached the maximum degradation rate at 331°C. The presence of SDS markedly reduced the activation energy in the degradation of SPI films. The FTIR spectra of gas products evolved during the thermal degradation indicated the formation of CO_2, CO, NH_3 and other unsaturated compounds, suggesting that the reaction mechanism included at the same time the scission of the C—N, C(O)—NH, C(O)—NH_2, NH_2 and C(O)—OH bonds of the protein.

The effect of stearic acid and glycerol on thermal properties of soy protein isolate (SPI) has been characterized [35]. TGA measurements showed that the thermal degradation of stearic acid modified SPI resin initiated at higher temperature than the SPI films. The stearic acid modified soy protein isolate resin began to degrade at 275°C as compared to glycerol plasticized resin at 250°C [35].

4.2. BIODEGRADABLE POLYMERS FROM PETROCHEMICAL SOURCES

4.2.1 Aliphatic polyesters and copolyesters

Thermal degradation of two aliphatic polyesters, poly(butylene succinate) (PBS) and poly (ethylene succinate) (PES), was studied by thermogravimetic analysis [36]. It was found that both polyesters exhibit a relatively good thermal stability since no significant weight loss occurred until 300°C. The decomposition step appeared at a temperature 399°C for PBS and 413°C for PES, indicating that PES is more stable than PBS. In both polyesters degradation takes place in two stages, the first corresponding to a very small mass loss, and the second at elevated temperatures, being the main degradation stage. It was supposed that the first decomposition step is due to oligomer degradation, whereas a random cleavage of the ester bond takes place, leading to the formation of carboxylic end groups and vinyl groups, as a predominant mechanism.

The strong influence of titanium alkoxide catalyst on thermal stability of poly(tetramethylene succinate) was found [37]. Physical properties of biodegradable aliphatic poly(butylene succinate-*co*-ethylene succinate) (PBES) and poly(butylene succinate-*co*-diethylene succinate) (PBDEGS) synthesized from succinic acid and 1,4-butanediol/diethylene glycol through a direct polycondensation were investigated [38]. TGA results indicated that thermal stabilities of the prepared PBDEGS significantly decreased with increasing the DEGS unit composition, while there was no marked decrease detected for those of the PBES copolyesters.

According to thermogravimetric measurements, aliphatic biodegradable polyester, poly (propylene succinate), shows a relatively good thermal stability since no significant weight loss occurred until 300°C in nitrogen [39].

4.2.2 Aromatic polyesters and copolyesters

The degradation of poly(trimethylene terephthalate) (PTT) with various molecular weights was studied in different atmospheres (argon, air and nitrogen) [40]. Under argon atmosphere PTT exhibited mainly one decomposition process, and all of the initial degradation temperatures, the temperature at the maximum weight-loss rate, the kinetic parameters for decomposition, increased with molecular weights. Under air atmosphere PTT exhibited two main degradation stages. The characteristic temperatures and kinetic parameters of thermal decomposition decrease from inert gas to air. High resolution thermogravimetry analysis in nitrogen at an auto-stepwise heating rate was applied to study the influence of molecular weight on thermal stability of poly(trimethylene terephthalate) [41]. The first small weight-loss stage (weight loss 2–4%) was highly sensitive to molecular weight. The temperature at the maximum weight-loss rate of this stage increased significantly with molecular weight. The weight loss during this step decreased steadily with increasing molecular weight, and was in good agreement with the value predicted on the basis of complete volatilization of the 1,3-propanediol unit and carbon dioxide devaluated from chain ends and the residual catalysts.

4.2.3 Poly(caprolactone) – PCL

A two-step degradation mechanism for polycaprolactone has been proposed by Persenaire *et al.* [42]. They studied thermal degradation of PCL by high resolution thermogravimetric analysis (TGA) simultaneously coupled with mass spectrometry (MS) and Fourier transform infrared spectrometry (FTIR). Based on evolved gas analysis by both MS and FTIR it was concluded that the first step was a random rupture of polyester chains via *cis*-elimination reaction which produced H_2O, CO_2, and 5-hexanoic acid. The second step is an unzipping depolymerization process at the chain ends with hydroxyl end groups to form ε-caprolactone (see Fig. 4.2).

The TGA and DTA studies reported by Aoyagi *et al.* [5] suggested a single-step degradation of PCL. However, they do not exclude the possibility of a random rupture of the polyester chain via a *cis*-elimination reaction, because it has been pointed out by Persenaire *et al.* [42] that the *cis*-elimination reaction and the unzipping depolymerization proceed consecutively at very close temperatures, so that these two steps may not be resolved by a conventional DTA technique.

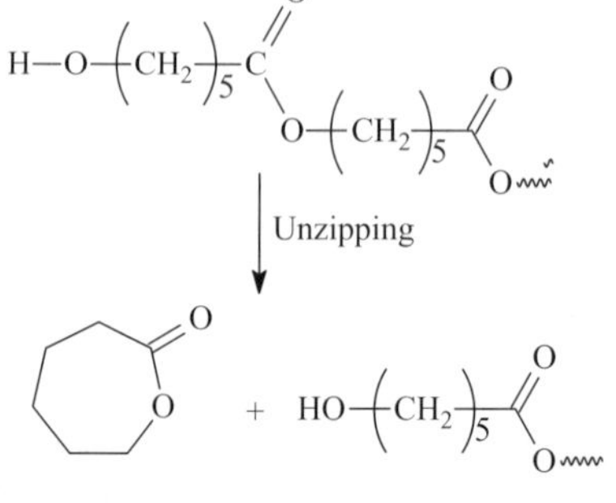

Figure 4-2 Unzipping depolymerization process during thermal degradation of PCL.

Recently, Sivalingham *et al.* [43] suggested that PCL underwent both random chain scission and specific chain end scission (elimination of monomer from the hydroxyl end of the polymer) simultaneously (a parallel mechanism) on non-isothermal heating and degraded by pure unzipping of the monomer from the hydroxyl end of the polymer chain on isothermal heating.

It was also reported that the thermal stability of the aliphatic polyesters, i.e. poly(ε-caprolactone) (PCL), poly(glycolide) (PGA), and poly(D,L-lactide) (PLA) investigated under dynamic heating in an inert atmopshere, based on the peak decomposition temperature, was in the order of PCL > PGA > PLA [44].

4.2.4 Poly(esteramides) – PEA

Thermal behaviour of laboratory synthesized poly(esteramide) constituted by the same monomers as commercial BAK 1095 (1,4-butanediol, adipic acid and 1,6-aminohexanoic acid), with a regular distribution of the monomers, was compared with the commercial reference polymer [45]. Both BAK 1095 and laboratory synthesized poly(esteramide) were thermally stable up to temperatures over 100°C higher than their melting temperatures (138°C and 159°C, respectively). Temperatures indicative of the beginning of the decomposition process and the 50% weight loss determined by thermogravimetry in nitrogen were 352 and 427°C for BAK 1095 polyesteramide, respectively [45]. Isothermal thermogravimetric analyis performed at 300°C indicated a weight loss close to 23% for a commercial BAK 1095 polyesteramide sample after 2 h.

4.2.5 Poly(vinyl alcohol) – PVA

The thermal degradation of poly(vinyl alcohol) (PVA) usually starts at about 150°C or above, depending on the PVA grade. The degradation process gives rise to the release of water from the polymer matrix, accompanied by the formation of volatile degradation products, such as acetic acid in partially acetylated samples [46].

The thermal degradation of poly(vinyl alcohol) was studied, both in the solid and molten state [47]. The thermal degradation of PVOH in the molten state consisted of water elimination and chain scission, via a six-member transition state, leading to the formation of volatile products including saturated and unsaturated aldehydes and ketones. In the solid state, thermal degradation of PVOH was exclusively by elimination of water.

4.3. BLENDS

The thermal stability of thermoplastic starch (TPS) blends containing various amounts of poly(hydroxybutyrate) (PHB) were evaluated [48]. Comparing PHB and TPS (soluble starch, glycerol 25%), the last one showed stability up to approximately 200°C at 10% of weight loss, while PHB exhibited the same weight loss only at 310°C. For blends containing a certain amount of PHB, thermal stability remains to a certain degree.

Thermal degradation of poly(lactic acid) (PLA) and poly(lactic acid)/corn starch composites with and without lysine diisocyanate (LDI) were evaluated by thermogravimetric analysis [49]. Thermal stability was decreased by addition of corn starch and the composites with lysine diisocyanate showed higher thermal degradation temperature than those without LDI.

Thermal degradation processes of series of mixtures based on polycaprolactone with microcrystalline cellulose and sisal fibre powder were investigated by thermogravimetric analysis, in order to predict the thermal behaviour of biodegradable matrixes reinforced with cellulose derivatives [50]. The stabilizing effect of microcrystalline cellulose on polycaprolactone degradation was observed. The same effect was observed for mixtures based on poly(3-hydroxybutyrate-8%-3-hydroxyvalerate) with microcrystalline cellulose [51]. Enhancement in thermal stability of both components was explained by strong hydrogen-type interactions.

4.4. SUMMARY OF THERMAL STABILITY OF COMPOSTABLE POLYMER MATERIALS

Apparent activation energy of thermal and thermooxidative degradation processes, as well as main decomposition temperatures in various environments of compostable polymer materials, are summarized in Tables 4.1 and 4.2, respectively.

Table 4.1. Apparent activation energy of degradation of compostable polymer materials

Polymer	Apparent activation energy, kJ/mol	Conditions	Ref.
Polymers from renewable sources			
PLA	43	Random chain scission	[44]
Poly(D,L-lactide)	105	Specific chain scission	
PLA (poly(L-lactic acid))	110	Under vacuum	[4]
PLA (poly(L-lactic acid))	92.6–105.3	In air	[54]
PLA (poly(L-lactic acid))	72–104	Under nitrogen; isothermal conditions	[55]
PDLA (poly(D-lactic acid))	155–242	Under nitrogen; isothermal conditions	[8]
PLLA (poly(L-lactic acid))	77–132	Under nitrogen; isothermal conditions	[8]
PLLA/PDLA (equimolar enantiomeric blend)	205–297	Under nitrogen; isothermal conditions	[8]
P3HB (poly(3-hydroxybutyrate))	212 ± 10	Random chain scission, 170–200°C; in nitrogen	[18]
PHB (poly(3-hydroxybutyrate))	235	Under vacuum	[4]
PHB (poly(3-hydroxybutyrate))	304.1	In air	[20]
P3HB3V (poly(3-hydroxybutyrate-*co*-3-hydroxyvalerate))	212 ± 10	Random chain scission, 170–200°C; in nitrogen	[18]
P3HB4V (poly(3-hydroxybutyrate-*co*-4-hydroxybutyrate))	212 ± 10	Random chain scission, 170–200°C; in nitrogen	[18]
PHBV (poly(3-hydroxybutyrate-*co*-3-hydroxyvalerate); 10.4 mol% HV)	325.4	In air	[20]
PHBV (poly(3-hydroxybutyrate-*co*-3-hydroxyvalerate); 20 mol% HV)	344.3–367.4	In air	[20]
PHB-HH (poly(3-hydroxybutyrate-*co*-3-hydroxyhexanoate))	189	Based on rheological tests	[56]
Starch	144.1 (corn) 171.6 (rice) 158.3 (potato) 159.3 (casava)	In nitrogen	[53]
Cellulose	172.1	In air	[27]

Table 4.1. (*Continued*)

Polymer	Apparent activation energy, kJ/mol	Conditions	Ref.
Cellulose acetate	83.9	In air	[27]
Cellulose propionate	133.0	In air	[27]
Cellulose	105.8	First DTG peak	[57]
derivative/starch blends	182.3	Second DTG peak	
(Mater-Bi)		In nitrogen	
Polymers from petroleum resources			
PBS	128	In nitrogen	[36]
(poly(butylene succinate))		First decomposition mechanism	
	189	Second decomposition mechanism	[36]
Poly(ethylene succinate)	182	In nitrogen	[36]
	256	First decomposition mechanism Second decomposition mechanism	
Poly(propylene succinate)	221	In nitrogen	[39]
Poly(ethylene adipate)	153	In nitrogen	[58]
Poly(propylene adipate)	121	First decomposition step	[58]
	152	Second decomposition step; in nitrogen	
Poly(butylene adipate)	185	First decomposition step	[58]
	217	Second decomposition step; in nitrogen	
PTT (poly(trimethylene terephthalate))	289*	First thermal degradation process; in nitrogen	[40]
	110*	in air	
PCL (poly(ε-caprolactone))	77	Random chain scission; in nitrogen	[44]
PVA (poly(vinyl alcohol))	150	In argon	[47]

* Average calculated from Freeman–Caroll, Friedman, and Chang techniques.

Table 4.2. Degradation temperatures of compostable polymer materials

Polymer	Decomposition temperature, °C	Remarks	Ref.
Polymers from renewable resources			
PLA	253.6	Onset temperature;	[1]
(poly(lactic acid))	339.5	Peak temperature; in helium	
PLA (poly(L-lactic acid))	360	T_{max}; under vacuum	[4]
PLLA (poly(L-lactic acid))	297	Starting temperature; in nitrogen	[8]
PDLA (poly(D-lactic acid))	287	Starting temperature; in nitrogen	[8]
PLLA/PDLA (equimolar enantiomeric blend)	292	Starting temperature; in nitrogen	[8]

(*Continued*)

Table 4.2. (*Continued*)

Polymer	Decomposition temperature, °C	Remarks	Ref.
PHB (poly(3-hydroxybutyrate))	290	T_{max}; under vacuum	[4]
PHB (poly(3-hydroxybutyrate))	349	Peak temperature; in nitrogen; dynamic conditions (high heating rate: 40°C/min)	[19]
PHB (poly(3-hydroxybutyrate))	246.3 268.0	The onset of degradation; T_{max} (temperature at the maximum decomposition rate) In air	[20]
PHBV (poly(3-hydroxybutyrate-*co*-3-hydroxyvalerate))	243.7 286.7	Onset temperature; Peak temperature; in helium	[1]
PHBV (poly(3-hydroxybutyrate-*co*-3-hydroxyvalerate))	352	Peak temperature; in nitrogen; dynamic conditions (high heating rate: 40°C/min)	[19]
PHBV (poly(3-hydroxybutyrate-*co*-3-hydroxyvalerate))	260.4 280.7	The onset of degradation; T_{max} (temperature at the maximum decomposition rate); in air	[20]
PHB-HH (poly(3-hydroxybutyrate-*co*-3-hydroxyhexanoate))	359	Peak temperature; under nitrogen; dynamic conditions (high heating rate: 40°C/min)	[19]
Starch-based blends (Mater-Bi ZF03U)	123.8 391.5	Onset temperature; Peak temperature; in helium	[1]
Starch-based blends (Mater-Bi Z101U/T)	56.1 410.8	Onset temperature; Peak temperature; in helium	[1]
Polymers from petroleum resources			
Poly(ethylene adipate)	379	Temperature at maximum decomposition rate; in nitrogen	[58]
Poly(propylene adipate)	385	Temperature at maximum decomposition rate; in nitrogen	[58]
Poly(butylene adipate)	412	Temperature at maximum decomposition rate; in nitrogen	[58]
Poly(ethylene succinate)	413	Temperature at maximum decomposition rate; in nitrogen	[36]
Poly(propylene succinate)	404	Temperature at maximum decomposition rate; in nitrogen	[39]
Poly(butylene succinate)	399	Temperature at maximum decomposition rate; in nitrogen	[36]
Poly(trimethylene terephthalate)	361–370	Temperature at maximum decomposition rate; different molecular weights; in argon	[41]
PCL (polycaprolactone)	363.3 413.6	Onset temperature; Peak temperature; in helium	[1]
PCL (polycaprolactone)	415	Maximum decomposition temperature; in nitrogen	[50]
PVA (poly(vinyl alcohol))	317	First decomposition peak; in nitrogen	[59]

REFERENCES

[1] Day M., Cooney J.D., Shaw K., Watts J.: Thermal analysis of some environmentally degradable polymers, *J. Therm. Anal.* 52 (1998) 261.

[2] Södergård A., Stolt M.: Properties of lactic acid based polymers and their correlation with composition, *Prog. Polym. Sci.* 27 (2002) 1123.

[3] Jamshidi K., Hyon S.-K, Ikada Y.: Thermal characterization of polylactides, *Polymer* 29 (1988) 2229.

[4] Kopinke F.-D., Remmler M., Mackenzie K., Mödr, Wachsen O.: Thermal decomposition of biodegradable polyesters – II. Poly(lactic acid), *Polym. Deg. Stab.* 53 (1996) 329.

[5] Aoyagi Y., Yamashita K., Doi Y.: Thermal degradation of poly[(R)-3-hydroxybutyrate], poly[ε-caprolactone], and poly[(S)-lactide], *Polym. Deg. Stab.* 76 (2002) 53.

[6] Cam D., Marucci M.: Influence of residual monomers and metals on poly(L-lactide) thermal stability, *Polymer* 38 (1997) 1879.

[7] Taubner V., Shishoo R.: Influence of processing parameters on the degradation of poly(L-lactide) during extrusion, *J. Appl. Polym. Sci.* 79 (2001) 2128.

[8] Tsuji H., Fukui I.: Enhanced thermal stability of poly(lactide)s in the melt by enantiometric polymer blending, *Polymer* 44 (2003) 2891.

[9] Bigg D.M.: Polylactide copolymers: effect of copolymer ratio and end capping on their properties, *Adv. Polym. Sci.* 24 (2005) 69.

[10] Yu H., Huang N., Wang Ch., Tang Z.: Modeling of poly(l-lactide) thermal degradation; theoretical prediction of molecular weight and polydispersity index, *J. Appl. Polym. Sci.* 88 (2003) 2557.

[11] Lenz R.W., Marchessault R.H. Bacterial polyesters: biosynthesis, biodegradable plastics and biotechnology, *Biomacromolecules* 6 (2005) 1.

[12] Kim K.J., Doi Y., Abe H.: Effects of residual metal compounds and chain-end structure on thermal degradation of poly(3-hydroxybutyric acid), *Polym. Deg. Stab.* 91 (2006) 769.

[13] Erceg M., Kovačić T., Klarić I.: Dynamic thermogravimetric degradation of poly(3-hydroxybutyrate)/aliphatic-aromatic copolyester blends, *Polym. Deg. Stab.* 90 (2005) 86.

[14] Grassie N., Murray E.J., Holmes P.A.: The thermal degradation of poly(D)-β-hydroxybutyric acid: Part 1 – Identification and quantitative analysis of products, *Polym. Deg. Stab.* 6 (1984) 47.

[15] Grassie N., Murray E.J., Holmes P.A.: The thermal degradation of poly(D)-β-hydroxybutyric acid: Part 2 – Changes in molecular weight, *Polym. Deg. Stab.* 6 (1984) 96.

[16] Grassie N., Murray E.J., Holmes P.A.: The thermal degradation of poly(D)-β-hydroxybutyric acid: Part 3 – The reaction mechanism, *Polym. Deg. Stab.* 6 (1984) 127.

[17] Griffin G.J.L., ed.: *Chemistry and Technology of Biodegradable Polymers*, Griffin, Chapman & Hall, Glasgow (1994), p.72.

[18] Kunioka M., Doi Y.: Thermal degradation of microbial copolyesters: poly(3-hydroxybutyrate-*co*-3-hydroxyvalerate) and poly(3-hydroxybutyrate-*co*-4-hydroxybutyrate), *Macromolecules* 23 (1990) 1933.

[19] He J.-D., Cheung M.K., Yu P.H., Chen G.-Q.: Thermal analyses of poly(3-hydroxybutyrate), poly (3-hydroxybutyrate-*co*-3-hydroxyvalerate), and poly(3-hydroxybutyrate-*co*-3-hydroxyhexanoate), *J. Appl. Polym. Sci.* 82 (2001) 90.

[20] Carrasco F., Dionisi D., Martinelli A., Majone M.: Thermal stability of polyhydroxyalkanoates, *J. Appl. Polym. Sci.* 100 (2006) 2111.

[21] Aggarval P., Dollimore D., Heon K.: Comparative thermal analysis study of two biopolymers, starch and cellulose, *J. Thermal Ana.* 50 (1997) 7.

[22] Soares R.M.D., Lima A.M.F., Oliveira R.V.B., Pires A.T.N., Soldi V.: Thermal degradation of biodegradable edible films based on xanthan and starches from different sources, *Polym. Deg. Stab.* 90 (2005) 449.

[23] Rudnik E., Matuschek G., Milanov N., Kettrup A.: Thermal properties of starch succinates, *Thermochim. Acta* 427/1–2 (2005) 163.

[24]   Rudnik E., Matuschek G., Milanov N., Kettrup A.: Thermal stability and degradation of starch derivatives, *J. Thermal. Anal. Cal.* 85 (2006) 267.

[25]   Rodriguez-Gonzalez F.J., Ramsay B.A., Favis B.D.: Rheological and thermal properties of thermoplastic starch with high glycerol content, *Carbohydr. Polym.* 58 (2004) 139.

[26]   Scheirs J., Camino G., Tumiatti W.: Overview of water evolution during the thermal degradation of cellulose, *Eur. Polym. J.* 37 (2001) 933.

[27]   Jain R.K., Lal K., Bhatnagar H.L.: Thermal, morphological, X-ray and spectroscopic studies on cellulose and its esters, *J. Appl. Polym. Sci.* 8 (1985) 359.

[28]   Jandura P., Riedl B., Kokta B.V.: Thermal degradation behavior of cellulose fibers partially esterified with some long chain organic acids, *Polym. Deg. Stab.* 70 (2000) 387.

[29]   Holme H.K., Foros H., Pettersen H., Dornish M., Smidsrød O.: Thermal depolymerization of chitosan chloride, *Carbohydr. Polym.* 46 (2001) 287.

[30]   Wanjun T., Cunxin W., Donghua C.: Kinetics studies on the pyrolysis of chitin and chitosan, *Polym. Deg. Stab.* 87 (2005) 389.

[31]   Pommet M., Morel M.-H., Redl A., Guilbert S.: Aggregation and degradation of plasticized wheat gluten during thermo-mechanical treatments, as monitored by rheological and biochemical changes, *Polymer* 45 (2004) 6853.

[32]   Wang S., Zhang S., Jane J.L., Sue H.J.: Effects of polyols on mechanical properties of soy protein. *Polym. Mater. Sci. Eng. (Am. Chem. Soc.)* 72 (1995) 88.

[33]   Kumar R., Choudhary V., Mishra S., Varma I.K., Mattiason B.: Adhesives and plastics based on soy protein products, *Ind. Crops Prod.* 16 (2002) 155.

[34]   Schmidt V., Giacomelli C., Soldi V.: Thermal stability of films formed by soy protein isolate-sodium dodecyl sulfate, *Polym. Deg. Stab.* 87 (2005) 25.

[35]   Lodha P., Netravali A.N.: Thermal and mechanical properties of environment – friendly "green" plastics from stearic acid modified–soy protein isolate, *Ind. Crops Prod.* 21 (2005) 49.

[36]   Chrissafis K., Paraskevopoulos K.M., Bikaris D.N.: Thermal degradation mechanism of poly(ethylene succinate) and poly(butylene succinate): comparative study, *Thermochim. Acta* 435 (2005) 142.

[37]   Yang J., Zhang S., Liu X., Cao A.: A study on biodegradable aliphatic poly(tetramethylene succinate): the catalyst dependences of polyester syntheses and their thermal stabilities, *Polym. Deg. Stab.* 81 (2003) 1.

[38]   Cao A., Okamura T., Nakayama K., Inoue Y., Masuda T.: Studies on syntheses and physical properties of biodegradable aliphatic poly(butylene-*co*-ethylene succinate)s, *Polym. Deg. Stab.* 78 (2002) 107.

[39]   Chrissafis K., Paraskevopoulos K.M., Bikaris D.N.: Thermal degradation kinetics of the biodegradable aliphatic polyester, poly(propylene succinate), *Polym. Deg. Stab.* 91 (2006) 60.

[40]   Wang X., Li X., Yan D.: Thermal decomposition kinetics of poly(trimethylene terephthalate), *Polym. Deg. Stab.* 69 (2000) 361.

[41]   Wang X., Li X., Yan D.: High-resolution thermogravimetric analysis of poly(trimethylene terephthalate) with different molecular weights, *Polymer Testing* 20 (2001) 491.

[42]   Persenaire O., Alexandre M., Degée P., Dubois P.: Mechanisms and kinetics of thermal degradation of poly(ε-caprolactone), *Biomacromolecules* 2 (2001) 288.

[43]   Sivalingam G., Karthik R., Madras G.: Kinetics of thermal degradation of poly(ε-caprolactone), *J. Anal. Appl. Pyrolysis* 70 (2003) 633.

[44]   Sivalingam G., Madras G.: Thermal degradation of binary physical mixtures and copolymers of poly(ε-caprolactone), poly(D,L-lactide), poly(glycolide), *Polym. Deg. Stab.* 84 (2004) 393.

[45]   Botines E., Rodríguez-Galán A., Puiggalí J.: Poly(ester amide)s derived from 1,4-butanediol, adipic acid and 1,6-aminohexanoic acid: characterization and degradation studies, *Polymer* 43 (2002) 6073.

[46]   Chiellini E., Corti A., D'Antone S., Solaro R.: Biodegradation of poly(vinyl alcohol) based materials, *Prog. Polym. Sci.* 28 (2003) 963.

[47] Holland B.J., Hay J.N.: The thermal degradation of poly(vinyl alcohol), *Polymer* 42 (2004) 6775.

[48] Lai S.-M., Don T.-M., Huang Y.-C.: Preparation and properties of biodegradable thermoplastic starch/poly(hydroxy butyrate) blends, *J. Appl. Polym. Sci.* 100 (2006) 2371.

[49] Ohkita T., Lee S.-H.: Thermal degradation and biodegradability of poly(lactic acid)/corn starch biocomposites, *J. Appl. Polym. Sci.* 100 (2006) 3009.

[50] Rusecksaite R., Jiménez A.: Thermal degradation of mixtures of polycaprolactone with cellulose derivatives, *Polym. Deg. Stab.* 81 (2003) 353.

[51] Fraga A., Rusecksaite R., Jiménez A.: Thermal degradation and pyrolysis of mixtures based on poly(3-hydroxybutyrate-8%-3-hydroxyvalerate) and cellulose derivatives, *Polym. Testing* 24 (2005) 526.

[52] Kopinke F.-D., Remmler M., Mackenzie K.: Thermal decomposition of biodegradable polyesters-I: poly(β-hydroxybutyric acid), *Polym. Deg. Stab.* 53 (1996) 25.

[53] Guinesi L.S., da Róz A.L., Corradini E., Mattoso L.H.C., de Teixeira E., da S. Curvelo A.A.: Kinetics of thermal degradation applied to starches from different botanical origins by non-iso-thermal procedures, *Thermochim. Acta* 447 (2006) 190.

[54] Gupta M.C., Deshmukh V.G.: Thermal oxidative degradation of poly-lactic acid, *Coll. Polym. Sci.* 260 (1982) 308.

[55] Babanalbandi A., Hill D.J.T., Hunter D.S., Kettle L.: Thermal stability of poly(lactic acid) before and after γ-radiolysis, *Polym. Int.* 48 (1999) 980.

[56] Daly P.A., Bruce D.A., Melik D.H., Harrison G.M.: Thermal degradation kinetics of poly(3-hydroxybutyrate-co-3-hydroxyhexanoate), *J. Appl. Polym. Sci.* 98 (2005) 66.

[57] Di Franco C.R., Cyras V.P., Busalmen J.P., Ruseckaite R.A., Vázquez A.: Degradation of polyc-aprolactone/starch blends and composites with sisal fibre, *Polym. Deg. Stab.* 86 (2004) 95.

[58] Zorba T., Chrissafis K., Paraskevopoulos K.M., Bikiaris D.N.: Synthesis, characterization and ther-mal degradation mechanism of three poly(alkylene adipate)s: comparative study, *Polym. Deg. Stab.* 92 (2007) 222.

[59] Cinelli P., Chiellini E., Gordon S.H., Imam S.H.: Characteristics and degradation of hybrid com-posite films prepared from PVA, starch and lignocellulosics, *Macromol. Symp.* 197 (2003) 143.

Chapter 5
Composting methods and legislation

Chapter 5
Composting methods and legislation

5.1. COMPOSTING DEFINITIONS

According to the US Environmental Protection Agency (EPA) composting means the controlled biological decomposition of organic material in the presence of air to form a humus-like material [1]. Controlled methods of composting include mechanical mixing and aerating, ventilating the materials by dropping them through a vertical series of aerated chambers, or placing the compost in piles out in the open air and mixing it or turning it periodically.

ASTM standards [2, 3] define composting as "a managed process that controls the biological decomposition and transformation of biodegradable materials into a humus-like substance called compost: the aerobic mesophilic and thermophilic degradation of organic matter to make compost; the transformation of biologically decomposable material through a controlled process of biooxidation that proceeds through mesophilic and thermophilic phases and results in the production of carbon dioxide, water, minerals and stabilized organic matter (compost or humus).

In the ISO draft standard [4] defines composting as "the autothermic and thermophilic biological decomposition of biowaste (organic waste) in the presence of oxygen and under controlled conditions by the action of micro- and macroorganisms in order to produce compost", where compost is defined as "organic soil conditioner obtained by biodegradation of a mixture consisting principally of vegetable residues, occasionally with other organic material and having a limited mineral content".

Other ISO standards [5–7] have more simplified definitions. Composting means an aerobic process designed to produce compost, while the definition of compost remains the same as in the above ISO draft [4].

British Standard (PAS 100) [8] defines composting as a process of controlled biological decomposition of biodegradable materials under managed conditions that are predominantly aerobic and that allow the development of thermophilic temperatures as a result of biologically produced heat, in order to achieve compost that is sanitary and stable.

According to British Standard (PAS 100) compost means solid particulate materials that are the result of composting, that have been sanitized and stabilized and that confer beneficial effects when added to soil and/or used in conjunction with plants.

5.2. COMPOSTING PROCESS AND METHODS

Composting is nature's way of recycling [9]. Composting decomposes and transforms organic material into a soil-like product called humus. The composting process uses microorganisms such as bacteria and fungi to break down the organic materials. For the process to work best, it is important that the microorganisms have a continuous supply of food (organics), water and oxygen. As well, managing the temperature of the composting materials is important to make the process work.

Compost can be made from most organic by-products [10]. Common feedstocks are poultry, hog, and cattle manures, food processing wastes, sewage sludge, municipal leaves, brush and grass clippings, sawdust, and other by-products of wood processing.

The main waste types that are composted include [11]:

- "green waste" (garden and park waste)
- "biowaste" (food waste)
- biodegradable waste streams from manufacturing (wood wastes, food processing wastes)
- mixed municipal solid waste
- sewage sludge
- slurries and manure from husbandry

During composting, microorganisms break down organic matter and produce carbon dioxide, water, heat and compost:

$$\text{Organic matter} + \text{microorganisms} + O_2 \text{ (air)} \rightarrow H_2O + CO_2 + \text{compost} + \text{heat}$$

Fresh compost is an intermediate product of the thermophilic stage, whereas mature compost is the end product of the stabilization stage [12]. Compost characteristics are essentially dependent upon the raw materials and the factors that affect the progress of the process.

Compost can be used in many applications depending on the quality produced and the quality of the product [9]. High quality compost is being used in agriculture, horticulture, landscaping and home gardening. Medium quality compost can be used in applications such as erosion control and roadside landscaping. Even low quality compost can be used as a landfill cover or in land reclamation projects.

Table 5.1. Common feedstocks and their characteristics [10]

Feedstock	Moisture content, %	Carbon:nitrogen ratio C:N
High in carbon		
Hay	8–10	15–30
Corn stalks	12	60–70
Straw	5–20	40–150
Corn silage	65–68	40
Autumn leaves	–	30–80
Sawdust	20–60	200–700
Brush, woodchips	–	100–500
Bark (paper mill waste)	–	100–130
Newspaper	3–8	400–800
Cardboard	8	500
Mixed paper	–	150–200
High in nitrogen		
Dairy manure	80	5–25
Poultry manure	20–40	5–15
Hog manure	65–80	10–20
Cull potatoes	70–80	18
Vegetable wastes	–	10–20
Coffee grounds	–	20
Grass clippings	–	15–25
Sewage sludge	–	9–25

There are many methods of composting organic materials and wastes, including three basic centralized types [9]:

- in-vessel method
- aerated static pile method
- windrow method

In the "in-vessel method", the organic material is composted inside a drum, silo, agitated bed, covered or open channel, batch container or other structure. The process conditions are closely monitored and controlled and the material is aerated and mechanically turned or agitated.

The "aerated static pile method" involves forming compostable materials into large piles, which are aerated by drawing air through the pile or forcing air out through the pile. The pile is not turned.

In the "windrow method", compostable material is formed into elongated piles, known as windrows, which are turned mechanically on a regular basis.

Composting systems comprise [13]:

- low-tech
  - windrow
- mid-tech
  - aerated static pile
  - aerated compost bins
- high-tech (in-vessel)
  - rotary drum composters
  - box/tunnel composting systems
  - mechanical compost bins

In-vessel composting systems are enclosed, rigid structures or vessels (reactors) used to contain the material undergoing biological processing [14]. They are equipped with process control systems that monitor the evolution of the biological activity by means of probes that measure the air temperature and the concentration of O_2 and/or CO_2. Monitoring the concentration of evolved gases enables precise determination of the status of the degradation process. In most plants, an air treatment unit is also included to limit the emission of particulate and gaseous pollutants into the atmosphere. In-vessel systems can be divided into two primary categories: vertical bioreactors and horizontal bioreactors. A vertical reactor is a cylindrical structure or container, composed of concrete or steel, and having a volume of a few hundred to more than $2000 \, m^3$ [14]. Typically, the material is loaded at the top and is extracted from the bottom in a continuous cycle. Aeration is carried out by forcing air from the bottom of the reactor by means of a centrifugal blower, countercurrent to the flow of the composting material. Vertical systems have been almost exclusively replaced by horizontal bioreactors. In horizontal systems, the biomass is maintained at the necessary aerobic conditions by means of forced aeration, usually combined with mechanical turning. The working cycle can be continuous or discontinuous.

Composting requires special conditions, particularly of temperature, moisture, aeration, pH and carbon to nitrogen (C/N) ratio, related to optimum biological activity in the various stages of the process [9].

Degradation of the waste in compost proceeds in three phases [9, 15]:

1. The first mesophilic phase
2. Thermophilic phase
3. Cooling and maturation phase

According to ASTM standard [3] the mesophilic phase is the phase of composting that occurs between 20° and 45°C, whereas the thermophilic phase means the phase in the composting process that occurs between 45° and 75°C; it is associated with specific colonies of microorganisms that accomplish a high rate of decomposition.

5.2.1 The first mesophilic phase [15]
At the beginning of composting, mesophilic bacteria and fungi degrade soluble and easily degradable compounds of organic matter, such as monosaccharides, starch, and lipids. Bacteria produce organic acids, and pH decreases to 5–5.5. Temperature starts to rise spontaneously as heat is released from exothermic degradation reactions. The degradation of proteins leads to release of ammonia, and pH rises rapidly to 8–9. This phase lasts from a few hours to a few days.

5.2.2 Thermophilic phase [15]
The compost enters the thermophilic phase when the temperature reaches 40°C. Thermophilic bacteria and fungi take over, and the degradation rate of the waste increases. If the temperature exceeds 55–60°C, microbial activity and diversity decrease dramatically. After peak heating, the pH stabilizes to a neutral level. The thermophilic phase can last from a few days to several months.

5.2.3 Cooling and maturation phase [15]
After the easily degradable carbon sources have been consumed, the compost starts to cool. After cooling, the compost is stable. Mesophilic bacteria and fungi reappear, and the maturation phase follows. However, most of the species are different from the species of the first mesophilic phase. Actinomycetes often grow extensively during this phase, and some protists and a wide range of macroorganisms are usually present. The biological processes are now slow, but the compost is further humified and becomes mature.

The duration of the phases depends on the composition of the organic matter and the efficiency of the process, which can be determined by oxygen consumption.

Microorganisms
Different communities of microorganisms predominate during the various composting phases [14–16]. Initial decomposition is carried out by mesophilic microorganisms, which rapidly break down the soluble, readily degradable compounds. The heat they produce causes the compost temperature to rise rapidly. As the temperature continues to rise, the bacterial growth rate slows down and subsequently the tolerance of the bacteria towards temperature is reduced. At 45°C, the mesophilic bacteria are inhibited, whereas thermophilic bacteria are activated [16]. The microbial populations during this phase are dominated by members of the genus *Bacillus*. At temperatures of 55°C and above, many microorganisms that are human or plant pathogens are destroyed. Because temperatures over about 65°C kill many forms of microbes and limit the rate of decomposition, compost managers use aeration and mixing to keep the temperature below this point.

Fungi, actinomycetes, and unicellular bacteria form the majority of compost microorganisms, and lagae, viruoses, protozoa, and macroorganisms make up the minority [15]. Bacteria are mostly heterotrophic, but denitrifying and nitrogen-fixing bacteria, hydrogen-oxidizing bacteria, and sulphur-oxidizing bacteria are also present. Actinomycetes often grow extensively during the cooling and maturation phase. Fungi grow in compost in all phases but may disappear temporarily during peak heating. A small but significant number of anaerobic bacteria have also been found in a compost environment. Anaerobic microenvironments may develop, especially during the thermophilic phase, when oxygen is rapidly consumed.

During the maturation phase protists and a wide range of macroorganisms, including mites, springtails, ants, millipedes, centipedes, spiders, beetles, and worms, appear in the compost.

Conditions of composting

The important parameters of composting are temperature, pH, moisture content, and oxygen transfer, which is regulated by aeration, free airspace, and agitation. The main properties of feed materials include C/N ratio, ratio size, rigidity, and nutrient and lignin compost.

Table 5.2. Optimal conditions for rapid, aerobic composting [10]

Conditions	Acceptable	Ideal
C:N ratios of combined feedstocks	20:1 to 40:1	25–35:1
Moisture content	40–65%	45–60%
Available oxygen concentration	>5%	>10% or more
pH	5.5–9.0	6.5–8.0
Temperature	43–66°C	54–60°C

Carbon and nitrogen are the two most important elements in the composting process, as one or the other is normally a limiting factor [18]. Carbon serves primarily as an energy source for microorganisms, while a small fraction of the carbon is incorporated in their cells. Nitrogen is critical for microbial population growth, as it is a constituent of protein which forms over 50% of dry bacterial cell mass. If nitrogen is limiting, microbial populations will remain small and it will take longer to decompose the available carbon. Excess nitrogen, beyond the microbial requirements, is often lost from the system as ammonia gas or other mobile nitrogen and can cause odours or other environmental problems. While the typically recommended C:N ratios for composting municipal solid waste (MSW) are 25:1 to 40:1 by weight, these ratios may need to be altered to compensate for varying degrees of biological availability.

Moisture management requires a balance between these two functions: microbial activity and oxygen supply [18]. Moisture is essential to the decomposition process, as most of the decomposition occurs in thin liquid films on the surfaces of particles. Excess moisture will fill many of the pores between particles with water, limiting oxygen transport. A minimum moisture content of 50 to 55% is usually recommended for high rate composting of MSW. The heat and airflow generated during composting evaporate significant amounts of water and tend to dry the material out. During the active composting phase, additional water usually needs to be added to prevent premature drying and incomplete stabilization. MSW compost mixtures usually start at about 52% moisture and dry to about 37% moisture prior to final screening and marketing.

Oxygen and temperature fluctuate in response to microbial activity, which consumes oxygen and generates heat [18]. Oxygen and temperature are linked by a common mechanism of control: aeration. Aeration both resupplies oxygen as it is depleted and carries away excess heat. Inadequate oxygen levels lead to the growth of anaerobic microorganisms which can produce odorous compounds. Oxygen concentrations in the large pores must normally be at least 12–14% (ideally 16–17%) to allow adequate diffusion into large particles and water filled pores. Most MSW composting systems used a forced aeration system with blowers and distribution pipes to supply oxygen during the initial phases of active composting.

Temperatures of 45 to 59°C provide the highest rate of decomposition, with temperatures above 59°C reducing the rate of decomposition due to a reduction in microbial diversity [18]. Since temperatures in excess of 55°C for several days are usually required for pathogen control, the ideal temperature operating range is relatively narrow. Composting systems attempt to control temperatures to a narrow range near 55 to 60°C in order to compromise between reaction rate, pathogen reduction, and odour generation. To maintain these temperature ranges, heat gains from microbial activity need to be balanced against heat losses, which occur primarily through evaporation of moisture and heating the aeration rate. Temperature, like oxygen supply, is usually managed by an aeration system: the same air which supplies oxygen can carry away excess heat.

Compost quality

Compost quality lies at the core of the issue of composting and biological treatment in general, as it defines the marketing potential and the outlets of the product and in most cases the viability of the treatment plant, but also the long-term acceptability of biological treatment as a valuable option in the waste hierarchy [19]. Compost quality refers to the overall state of the compost in regard to physical, chemical and biological characteristics, which indicate the ultimate impact of the compost on the environment. The criteria that are relevant to the evolution of quality depend on what purpose the compost is used for, the relevant environmental protection policies and the market requirements. For example, compost intended as growing media should meet more stringent quality criteria compared to composts that will be used as landfill cover.

Compost quality criteria include a variety of parameters, such as particle size distribution, moisture, organic matter and carbon content, concentration and composition of humus-like substances, nitrogen content and forms of nitrogen, phosphorus and potassium, heavy metals, salinity and the nature of ions responsible for it, cation exchange capacity, water holding capacity, porosity and bulk density, inert contaminants, pathogens, and state of maturity and stability [19, 20]. However, the most important, from the point of view of standards for the protection of public health, the soil and the environment in general, are those relating to pathogens, inorganic and organic potentially toxic compounds (heavy metals, PCBs, PAHs, phthalates, etc.) and stability, the latter determining compost nuisance potential, nitrogen immobilization and leaching and phytotoxicity [19]. Pathogen levels and trace metal concentrations are strictly regulated for biosolid compost. Compost stability and maturity indexes, however, are generally not regulated.

Stability and maturity are terms often used to characterize compost, yet opinions about what these terms mean vary widely. Compost stability refers to the resistance of compost organic matter to further rapid degradation, and can be directly measured by respirometric rates [20]. Compost maturity is related to suitability for plant growth, although some authors also relate it to humification [20]. The Waste and Resources Action programme report [21] defines stability as "the rate of biological activity" that is measured as the rate of aerobic respiration using a standardized CO_2 evolution. Maturity is defined as "readiness for use" and assessed by stability and phytotoxicity, plus other direct parameters relevant to the intended use.

In the United States, there is no one standard approach to assessing stability. In recent work by the California Compost Quality Council (CCQC) in conjunction with the California Integrated Waste Management Board (CIWMB), Woods End Laboratory and other peer-reviewers, maturity has been defined as the degree of completeness of composting [22]. This is in contrast to earlier definitions used in America and indicates that maturity is no longer viewed as a single property that can be tested for separately. Maturity must be assessed by measuring two or more parameters of compost, after the C:N ratio has been measured.

The maturity of compost is important for application purposes [23]. During the last two decades scientists have been looking for reliable parameters to determine compost maturity, for example plant growth, respiration rates, humification index, and water-soluble C/total N ratio [20]. However, it is unlikely that one single parameter will be found to assess compost maturity, mainly because of the great variety of composting feedstocks and management practices [20]. As a measure of stability three methods are commonly used: calorimetry (Dewar self-heating), oxygen demand and CO_2 evolution. Commercial test kits are available for routine testing of compost and maturity, e.g. Solvita test. The Solvita test is a rapid, colorimetric procedure which measures the CO_2 respiration and ammonia (NH_3) evolution in a specified volume of compost. Usually, the degree of compost maturity is calculated by the maximum self-heating temperature measured in an isolated vessel [23]. Internationally recognized and relatively easy to conduct is the "Dewar-flask test". In a Dewar-flask test compost is placed in a 2 litre insulated flask and the temperature of the product compared with ambient. The temperature of the compost in the flask must not increase by more than 10°C above ambient. Another method – accepted and standardized as well – is the measurement of respiration activity in the respirometer over a period of four days.

This method requires a considerable lesser amount of material (20 g, compared to 2 l for the self-heating test). The maturity is calculated as described in Table 5.3, according to Jourdan and Becker methods [23].

Table 5.3. The relation between the degree of maturity and some biochemical parameters [23]

Degree of maturity	Maximum temperature, °C	O_2 consumption (mg/g OS) according to Jourdan	O_2 consumption (mg/g OS) according to Becker	Material status
I	>60	>40	>80	Raw material
II	60–50.1	40–8.1	80–50	Fresh compost
III	50–40.1	28–16.1	50–30	Fresh compost
IV	40–30.1	16–6.1	30–20	Maturated compost
V	≤30	≤6	≤20	Maturated compost

Compost quality standards

Compost quality guidelines are relatively new, dating to the mid-1980s [24]. There has been a steady progression of definitions of contaminant limits when considering compost quality. The very first published limits pertained to heavy metals and were seen in the late 1970s in Europe. These standards include:

1. Heavy metal allowable levels
2. Physical contamination and inert contamination
3. Pathogenic bacteriology and phytopathogens
4. Potentially toxic elements (PTEs)
5. Maturity and plant growth performance

Currently, compulsatory and voluntary compost standards in different countries are characterized by a great degree of heterogeneity, stemming from the effort to combine two often contradicting targets: maximum environmental and public health protection on the one hand and maximum organic matter recycling on the other. Moreover, the precautionary approach adopted in the EU and the risk assessment approach prevailing in the USA, may lead to broad differences in the accepted limit values for a number of critical parameters, such as heavy metals [19, 22]. The quality criteria upon which compost standards are based vary across the countries in the range of criteria, the requirements, and the limit values.

Of all potential quality standards, heavy metals have been the focus of most attention. Permissible metal ranges reveal significant variation within Europe [22]. However, United States numbers diverge dramatically with regard to allowed Cd, Cr, Cu, Hg and Ni (Table 5.4).

Even within the EU there is a wide variation among the limit values adopted by the member countries, with the north being usually more stringent than the south, reflecting mainly the varying level of progress on source separation of the biodegradable fraction of MSW, but also the different needs in soil organic matter [19].

All national compost standards include compost sanitization criteria for human pathogens and occasionally for plant pathogens [19]. These criteria may refer to the product (absence of *Salmonella*, absence or low levels of faecal coliforms and faecal streptococci), the process (setting a minimum period for which the compost should maintain a temperature higher than a designated level) or both maximum permissible values are set for heavy metals (Cd, Cr, Cu, Hg, Ni, Pb, Zn) although the limits vary widely. Similar values are set for foreign matter (glass, plastics and stones) in most national specifications, usually defined as maximum allowed content on a dry weight basis and with reference to their particle size. The degree of compost stability and its nitrogen content are particularly important for its agronomic use and are increasingly more often defined in compost specifications.

The British Standards Institution Publicly Available Specification (PAS) 100 [8] covers the range of materials used to make the compost, their quality and traceability, the minimum requirements for the process of composting and the quality of the end product. The PAS 100 specification is a minimum specification and limits stones, weed seeds and physical and chemical contaminants.

Compost use
Compost is used as (organic) fertilizer, soil improver/conditioner, manufactured topsoil, growing medium, and mulch for use in [25]:

- Agriculture (intensive, organic)
- Fruit growing and wine making

Table 5.4. Comparison of heavy metals limits in EC states versus USA [22]

Metal	Symbol	EU range, mg/kg	USA biosolids, mg/kg
Cadmium	Cd	0.7–10	39
Chromium	Cr	70–200	1200
Copper	Cu	70–600	1500
Mercury	Hg	0.7–10	17
Nickel	Ni	20–200	420
Lead	Pb	70–1000	300
Zinc	Zn	210–4000	2800

Table 5.5. Minimum compost quality for general use according to BSI PAS 100 [8]

Parameter	Test method	Unit	Upper limit
Pathogens (human and animal indicator species)			
Salmonella spp.	ABPR 2003, schedule 2, part II or BS EN ISO 6579	25 g fresh mass	Absent
Escherichia coli	BSI ISO 11866-3	$CFU g^{-1}$ fresh mass	1000
Potentially toxic elements			
Cadmium (Cd)	BS EN 13650 (soluble in aqua regia)	$mg kg^{-1}$ dry matter	1.5
Chromium (Cr)	BS EN 13650 (soluble in aqua regia)	$mg kg^{-1}$ dry matter	100
Copper (Cu)	BS EN 13650 (soluble in aqua regia)	$mg kg^{-1}$ dry matter	200
Lead (Pb)	BS EN 13650 (soluble in aqua regia)	$mg kg^{-1}$ dry matter	200
Mercury (Hg)	BS ISO 16772	$mg kg^{-1}$ dry matter	1.0
Nickel (Ni)	BS EN 13650 (soluble in aqua regia)	$mg kg^{-1}$ dry matter	50
Zinc (Zn)	BS EN 13650 (soluble in aqua regia)	$mg kg^{-1}$ dry matter	400
Stability/maturity			
Microbial respiration rate	ORG0020	$mg CO_2$/g organic matter/day	16
Plant response			
Germination and growth test	BSI PAS 100: 2005, Annex D	Reduction in germination of plants in amended compost as % of germinated plants in peat control	20
		Reduction of plant mass above surface in amended compost as % of plant mass above surface in peat control	20
		Description of any visible abnormalities	No abnormalities
Weed seeds and propagules			
Germinating weed seeds or propagules regrowth	BSI PAS 100: 2005, Annex D	Mean number per litre of compost	0
Physical contaminants			
Total glass, plastic, and any "other" non-stone fragments >2 mm	BSI PAS 100: 2005, Annex E	% mass/mass of "air-dry" sample	0.5 (of which 0.25 is plastic)
Stones			
Stones >4 mm in grades other than "mulch"	BSI PAS 100: 2005, Annex E	% mass/mass of "air-dry" sample	8
Stones >4 mm in "mulch" grade	BSI PAS 100: 2005, Annex E	% mass/mass of "air-dry" sample	16

- Horticulture
- Potting
- Nurseries
- Greenhouses
- Private gardens
- Landscaping (e.g. parks)
- Ground rehabilitation
- Silviculture

5.3. COMPOSTING OF BIODEGRADABLE POLYMERS

5.3.1 Composting standards

In Europe the origin for composting standards is related to the European Directive on Packaging and Packaging Waste [26–28]. According to Directive the European Commission shall promote, in particular, the preparation of European standards relating among other things to criteria for composting methods and produced compost and criteria for the marking of packaging. It is noteworthy that Annex II of the Directive defines in general the criteria for packaging recoverable in the form of composting as follows: packaging waste processed for the purpose of composting shall be of such a biodegradable nature that it should not hinder the separate collection and the composting process or activity into which it is produced.

5.3.2 Biodegradable packaging

Biodegradable packaging waste shall be of such a nature that it is capable of undergoing physical, chemical, thermal or biological decomposition such that most of the finished compost ultimately decomposes into carbon dioxide, biomass and water.

In particular, the emphasis is given that recovery and recycling of packaging waste should be further increased to reduce its environmental impact.

5.3.3 Standards relating to specification for compostable plastics

- ASTM D 6400 [2]
- ISO/DIS 17088 [4]
- EN 13432 [29]
- DIN-54900 [30]

ASTM 6400-04 Standard specifications for compostable plastics [2]

For a plastic to be claimed biodegradable under composting conditions or compostable, it has to meet the Specifications Standards ASTM D 6400. This specification is intended to establish the requirements for labelling of materials and products, including packaging made from plastics, as "compostable in municipal and industrial composting facilities".

The key requirements include:

- Mineralization (conversion to carbon dioxide, water and biomass via microbial assimilation at the same rate as natural materials (leaves, paper, grass and food scraps))
- Disintegration
- Safety

The properties in this specification are those required to determine if plastics and products made from plastics will compost satisfactorily, including biodegrading at a rate comparable to known compostable materials. Further, the properties in the specification are required to assure that the degradation of these materials will not diminish the value or utility of the compost resulting from the composting process.

According to ASTM requirements 90% of the carbon of the test materials must be assimilated by the compost microorganisms as documented by measuring CO_2 production, within a six month period, extendable to one year if radiolabelled carbon is used. Moreover, disintegration of the film or article of the use thickness such that less than 10% of the material remains on a 10 mesh screen after sieving must be proved. Safety of compost must be proved by testing phyto- or ecotoxicity using methods listed in the Standard.

ISO/DIS 17088 Specifications for compostable plastics [4]
This International Standard specifies test methods and requirements to determine and to label plastic products and products made from plastics that are designed to be recovered through aerobic composting by addressing four characteristics:

1. Biodegradation
2. Disintegration during biological treatment
3. Negative impacts upon the biological treatment process and facility
4. Negative impacts on the quality of the resulting composts, including the presence of restricted metals and other harmful ingredients

This specification is intended to establish the requirements for labelling of materials and products, including packaging made from plastics, as "compostable", "compostable in municipal and industrial composting facilities" and "biodegradable during composting". Manufacturers shall conform to all international, national, local or regional regulations when labelling these products (e.g. European Directive 94/62/EC).

Recovery of compostable plastics through composting can be obtained under the environmental conditions found in well-managed composting plants where thermophilic conditions, water content, aerobic conditions, the carbon/nitrogen ratio and processing conditions are optimized. This is generally obtained in industrial and municipal composting plants. Under these conditions, "compostable plastics" will disintegrate and biodegrade at rates comparable to yard trimmings, kraft paper bags and food scraps.

EN 13432:2000 Packaging – Requirements for packaging recoverable through composting and biodegradation – Test scheme and evaluation criteria for the final acceptance of packaging [29]
EN 13432 has been published in the *Official Journal of the European Communities*, to become a harmonized norm. It became a tool to prove compliance with European Directive 94/62/EC, and is recognized both at European level in each Member State and by the International Standards Organization. The Directive on Packaging and Packaging Waste (94/62/EC) defines requirements for packaging to be considered recoverable. EN 13432 amplifies these requirements with respect to organic recovery. Organic recovery of packaging and packaging materials, which includes aerobic composting and anaerobic biogasification of packaging in municipal or industrial biological waste treatment facilities, is an option for reducing and recycling packaging waste. Thus, using these biological technologies, the aims of Directive 94/62/EC can be met.

EN 13432 specifies requirements and procedures to determine the compostability and anaer-obic treatability of packaging and packaging materials by addressing four characteristics:

1. Biodegradability
2. Disintegration during biological treatment
3. Effect on the biological treatment
4. Effect on the quality of the resulting compost

In the case of packaging formed by different components, some of which are compostable, the packaging itself as a whole is not compostable. However, if the components can be easily separated by hand and before disposal, the compostable components can be effectively consid-ered and treated as such, once separated from the non-compostable components.

EN 13432 defines the characteristics a material must own in order to be claimed as "compostable" and, therefore, recycled through composting of organic solid waste. According to EN 13432, the compostability criteria include:

- Biodegradability, namely the capability of the compostable material to be converted into CO_2 under the action of microorganisms. The ISO 14855 standard is recommended as a laboratory test method, i.e. determination of the ultimate aerobic biodegradability and dis-integration of plastic materials under controlled composting conditions. ISO 14851 and ISO 14852 standards can also be used. In order to show complete biodegradability, a biodegrad-ation level of at least 90% must be reached in less than six months.
- Disintegrability, namely fragmentation and loss of visibility in the final compost (absence of visual pollution). The standard recommends assessment of disintegration through trials on a pilot or full-scale composting plant. Specimens on the test materials are composted with biowaste for three months. The final compost is then screened with a 2 mm sieve. The mass of test material residues with dimension >2 mm shall be less than 10% of the original mass.
- Absence of negative effects on the composting systems. Introduction of the packaging waste should not have a negative impact on the operation of the plant. Verified with the pilot scale composting test.
- Absence of negative effects on the final compost. The compost samples generated in tri-als should be compared against control samples taken from the same process without packaging waste feedstock. The results must be comparable, as well as conforming with European and national standards. Physicochemical parameters should include volumetric weight/density, total dry solids, volatile solids, salt content, pH and presence of N_2, NH_3, P, Mg and K. Ecotoxicological effects on the plant growth should be determined and a plant growth test (modified OECD 208) applied.

Moreover, evaluation criteria of the packaging, packaging material or packaging component to be claimed compostable include:

- Volatile solid contents (at least 50%)
- Low levels of heavy metals (below given max. values, cf. Table 5.6)

Table 5.6. Comparison of concentrations of regulated heavy metals in different countries [4]

Element mg/kg on dry substance	US ASTM D 6400	Canada ASTM D 6400	European Union EN 13432	Japan
Zn	1400	925	150	180
Cu	750		50	60
Ni	210	90	25	30
Cd	17	10	0.5	0.5
Pb	150	250	50	10
Hg	8.5	2.5	0.5	0.2
Cr			50	50
Mo		10	1	
Se	50	7	0.75	
As	20.5	37.5	5	5
F			100	
Co		75		

Biodegradation standards and ecotoxicological assessment of compostable polymer materials are described in detail in Chapters 6 and 7, respectively.

5.3.4 Comparison of standards

The ASTM D 6400 and ISO/DIS 17088 standards all define biodegradability in respect of a time period of 180 days. In the case of EN 13432 a material is deemed biodegradable if it will break down to the extent of at least 90% to H_2O and CO_2 and biomass within a period of six months. Each of the named standards sets limits for the amounts of heavy metals that the material may contain. German standard DIN V 54900 sets the strictest standards, that is, it permits the lowest value of heavy metal presence. DIN V 54900 is the oldest of its kind and still has some relevance in Germany. However, it was replaced by the European EN 13432 standard. A number of publications concerning compostability of polymers, especially ecotoxicity tests, have been based on German DIN V 54900 (cf. Chapter 7).

DIN V 54900 consists of four parts:

1. DIN V 54900-1 Testing of compostability of plastics – Part 1: Chemical testing, October 1998
2. DIN V 54900-2 Testing of compostability of plastics – Part 2: Testing of the complete biodegradability of plastics in laboratory tests, September 1998
3. DIN V 54900-3 Testing of compostability of plastics – Part 3: Testing under practice-relevant conditions and a method of testing the quality of the composts, September 1998
4. DIN V 54900-4 Testing of compostability of plastics – Part 4: Testing of ecotoxicity, January 1997

EN 13432 is a harmonized, mandated European standard, which is currently the most relevant standard in Europe. It is valid in all EU Member States.

In general, the concept of compostability, i.e. criteria in all standards, is similar (Fig. 5.1).

Table 5.7. Comparison of key requirements of composting standards

Standard	Biodegradation	Disintegration	Safety
ASTM D 6400	• For products consisting of a single polymer (homopolymer or random copolymer), 60% of the organic carbon must be converted to carbon dioxide within 180 days • For products consisting of more than one polymer (block copolymers, segmented copolymers, blends or addition of low molecular weight), 90% of the organic carbon must be converted to carbon dioxide within 180 days	No more than 10% of its original dry weight remains after sieving on a 2.0 mm sieve after controlled laboratory scale composting	• No adverse impact on ability of compost to support plant growth • Low levels of heavy metals
ISO/DIS 117088	• For products consisting of a homopolymer, 60% of the organic carbon must be converted to carbon dioxide within 180 days • For all other polymers (e.g. copolymers or blends), 90% of the organic carbon must be converted to carbon dioxide within 180 days	No more than 10% of its original dry mass remains after sieving on a 2.0 mm sieve after 84 days in a controlled composting test	• Low levels of heavy metals • A minimum of 50% of volatile solids • Ecotoxicological assessment (plant growth test on two different plant species following modified OECD guideline 208)
EN 13432	At least 90% of biodegradation within six months	No more than 10% of the residues from the packaging waste should be larger than 2 mm	• Low levels of heavy metals • Physical/chemical analysis of the resulting compost • Ecotoxicological assessment (plant growth test on two different plant species following modified OECD guideline 208)

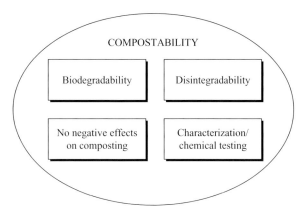

Figure 5-1 The criteria of compostability.

5.4. LABELLING SYSTEMS IN DIFFERENT REGIONS

Labelling serves to identify and verify the compostability of a product. The compostable logos are designed to address the confusion that has existed by building credibility and recognition for products that meet the compostability standards among consumers, waste management regulators and others.

Labelling systems for compostable polymer materials exist in Europe, the USA and Japan.

5.4.1 In Europe
1. The compostability mark of European Bioplastics and DIN CERTCO:

Compostable

DIN CERTCO operates a certification scheme for compostable products made of biodegradable materials and licenses the use of the corresponding mark developed by European Bioplastics (the former IBAW Interest Group for Biodegradable Materials).

Certification organisation: DIN CERTCO (Germany)

Website: www.dincertco.de

Requirements: The certification procedure for products made of compostable materials may be performed on the basis on the following standards:

- DIN V 54900 "Testing of the compostability of plastics" (replaced by DIN EN 13432 – after the withdrawal of this standard in February 2004, the testing basis is only used for those product certifications where the basic material according to this standard is registered by DIN CERTCO)

- DIN EN 13432 "Packaging – Requirements for packaging recoverable through composting and biodegradation – Test scheme and evaluation criteria for the final acceptance of packaging"
- ASTM D 6400 "Standard specification for compostable plastics"

The tests are to be conducted consistently in accordance with one of the three standards, i.e. the chemical test, the test for ultimate biodegradability and the test for compostability under practice-relevant conditions and of the quality of the composts.

Chemical testing serves to ensure that neither harmful organic substances, such as polychlorinated biphenyl (PCB) and dioxins, nor heavy metals, such as lead, mercury and cadmium, pass into the soil via the compost.

The method specified for the testing of biodegradability and of disintegration serve to verify the complete degradation of the materials within the processing period of normal composting plants. Testing of ultimate biodegradation is conducted in accordance with one of the following methods:

1. As specified in DIN V 54900
   - Method 1: Measuring the oxygen demand in a closed respirometer
   - Method 2: Analysis of evolved carbon dioxide in an aqueous medium
   - Method 3: Analysis of evolved carbon dioxide in compost
2. As specified in DIN EN 13432
   - ISO 14855 "Determination of the ultimate aerobic biodegradability and disintegration of plastic materials under controlled composting conditions – method by analysis of evolved carbon dioxide"
   - DIN EN 14046 "Packaging – Evaluation of the ultimate aerobic biodegradability of packaging materials under controlled composting conditions – Method by analysis of released carbon dioxide"
   - ISO 14851 "Determination of the ultimate aerobic biodegradability of plastic mater-ials in an aqueous medium – Method by measuring the oxygen demand in a closed respirometer"
   - ISO 14852 "Determination of the ultimate aerobic biodegradability of plastics materials in an aqueous medium – Method by analysis of evolved carbon dioxide"
3. As specified in ASTM D 6400
   - ASTM D 5338-98 "Standard test method for determining aerobic biodegradation of plastics materials under controlled composting conditions"
   - ASTM D 6002-96 "Standard guide for assessing the compostability of environmentally degradable plastics"

Testing of compostability under practice-relevant conditions is conducted in accordance with the following methods:

1. As specified in DIN V 54900
   - Testing in a pilot-scale test under optimized process conditions (with determination of maximum compostable material thickness)
   - Testing in a composting plant under real conditions (with determination of maximum compostable material thickness)
2. As specified in DIN EN 13432
   - Testing in a pilot-scale test
   - Testing in a composting plant under real conditions

3. As specified in ASTM D 6400
    *   Method specified in subclause 7.2.1 of ASTM D 6002-96 "Standards guide for assessing the compostability of environmentally degradable plastics"

An ecological non-toxicity test which is also prescribed ensures that the plastics used have no adverse effect on the quality of the compost. Additionally the maximum compostable layer thickness is determined.

If the results of the tests are in conformity with the standard(s) and/or the certification scheme, the material, intermediate or additive, is registered and included in a positive list.

Verification tests are performed in order to verify that the same base materials as those declared on application for certification are being used. For this purpose, infrared spectra are recorded and compared.

The compostability mark is accepted in Germany, Switzerland, the Netherlands, the United Kingdom, and in Poland. Certificates for compostable materials and products are valid in all these countries. Additional tests are not necessary.

Products that fulfil the requirements: Products of 24 producers are certified by DIN CERTCO as compostable materials. Examples of polymeric materials recognized and labelled as compostable include: polylactic acid (e.g. Nature Works PLA, Mitsui Lacea), thermoplastic starch (BIOTEC BIOPLAST), polycaprolacone (Solvay Caprolactones Solvay Interox CAPA), starch-based blends (Novamont Mater Bi), aliphatic–aromatic copolyesters (BASF Ecoflex, DuPont Biomax).

2. The compostability mark of AIB Vinçotte (Belgium):

Certification organization: AIB Vinçotte (Belgium) is the European international testing and certification company based in Brussels. The OK compost mark guarantees that a material of a product can be composted in an industrial plant.

Website: www.vincotte.com

Requirements: Testing according to European standard: EN 13432.

Products that fulfil the requirements: Examples of polymer materials that obtained the OK compost mark include: blends of aliphatic–aromatic copolyesters and polylactic acid (BASF Ecovio), starch based (Biotec BIOPLAST), starch-based blends (Novamont Mater Bi), aliphatic copolyesters (Showa High Polymer Bionolle).

3. Compostability mark of Finnish Solid Waste Association (Jätelaitosyhdistys) (Finland)

Certification organization: Finnish Solid Waste Association, FSWA (Jätelaitosyhdistys) (Finland) represents Finnish regional and municipal waste management companies. The FSWA is a member of the ISWA (International Solid Waste Association) – federation of organizations, local authorities, and private companies in the field of waste management. About ten years ago, FSWA started to collect biowaste, mainly organic kitchen waste, from households. Part of a large communication and promotion campaign, in which the Ministry for Environment was also involved, was the creation of the "Finnish apple logo". In order to distinguish compostable biowaste bags from "normal" plastic bags, the apple logo was printed on the biowaste bags. In Finland, biowaste bags are more or less the only compostable plastics products that are available. All biowaste bags carrying the apple logo are certified according to EN 13432:2000. They must fulfil requirements of compostability, biodegradability and ecotoxicity. Laboratory test results from a reliable Finnish or European laboratory are required.

Website: www.jatelaitosyhdistys.fi

Requirements: Testing according to European standard: EN 13432.

Products that fulfil the requirements: In Finland today about five suppliers sell certified biowaste bags with the apple logo.

5.4.2 In America

The logo has been developed jointly by the international Biodegradable Products Institute (BPI) – a government/industry/academic association that promotes the use of biodegradable polymer materials – and the US Composting Council (USCC) – representing the composting industry in the USA.

Website: www.bpiworld.org

Certification organization: Biodegradable Products Institute/US Composting Council (USA).

Requirements: Testing according to standard: ASTM 6400.

Products that fulfil the requirements: Products that satisfied the BPI-USCC compost label requirements include: starch-based blends (Novamont SPA. Mater Bi™), aliphatic–aromatic copolyesters (Eastman Eastar Bio™, BASF Ecoflex), polylactic acid (Cargill Dow PLA NatureWorks™), starch and polylactic blends (Cereplast Cereplast™).

5.4.3 In Asia

In Japan the Biodegradable Plastics Society (BPS), an industry group, established the GreenPla identification system in June 2000 based on examinations of the experiment and assessment methods for biodegradable plastics and their safety that METI (then the Ministry of International Trade and Industry) carried out from 1989 to 1999 [31]. GreenPla is nickname for biodegradable plastics. Ministry of International Trade and Industry, Japan Bioindustry Association (JBA) and BPS jointly called for applications for nicknames. GreenPla won the Minister of International Trade and Industry prize. Plastic products composed only of material whose safety and biodegradability has been confirmed are certified as GreenPla products and efforts are being made to differentiate these products from other plastic products by using the unified symbol mark. Starting in 2002, the BPS established and began administering standards on the compostability of certified products. By the end of 2005, over 800 products had obtained the GreenPla mark.

Certification organization: Biodegradable Plastic Society (Japan).

Website: www.bpsweb.net/english

Testing according to standard: GreenPla certification programme.

Requirements:

- Identification of all constituent substances (components) of product
- All components are listed on the Positive List and are to comply with the Rules for Positive List (Table 5.8)
- To include 50.0 wt% or more, or 50.0 vol% or more of organic substances in a product
- Amounts of specified elements included in a product are not to exceed the upper limits (Table 5.9)

Products that fulfil the requirements: Polymers that fulfil the requirements of the certification system for the GreenPLA logo include: poly(hydrohybutyrate) (PHB), poly(lactic acid)

Table 5.8. Rules for Positive List for GreenPla products

I. Biodegradability tests	OECD 301C (Modified MITI test – Ready degradability test) Chemical substance – Aerobic biodegradability test by activated sludge
	JIS K 6950 (ISO 14851)
	Plastics – Evaluation of ultimate aerobic biodegradability in an aqueous medium (method by determining the oxygen demand in a closed respirometer)
	JIS K 6951 (ISO 14852) Plastics – Evaluation of ultimate aerobic biodegradability in an aqueous medium (method by analysis of released carbon dioxide)
	JIS K 6953 (ISO 14855)
	Plastics – Evaluation of the ultimate aerobic biodegradability and disintegration under the controlled composting conditions (method by analysis of released carbon dioxide)
II. Oral acute toxicity test	Test on rats
III. Environmental safety tests	OECD test guideline 201 Algae, growth inhibitor test OECD test guideline 202 *Dapnia* sp., reproduction test at 14 days OECD test guideline 203 Fish, acute toxicity test

Table 5.9. Upper limit of contents of specified elements in GreenPla product

Name of element	Upper limit, ppm
Cadmium (Cd)	0.5
Lead (Pb)	50.0
Chromium (Cr)	50.0
Arsenic (As)	3.5
Mercury (Hg)	0.5
Copper (Cu)	37.5
Selenium (Se)	0.75
Nickel (Ni)	25.0
Zinc (Zn)	150.0
Molybdenum (Mo)	1.0
Fluorine (F)	100.0

Table 5.10. International Compostability Certification Network

Organization	Region/Country	Cooperation
DIN CERTCO	Europe	BPS, BPI
Biodegradable Products Institute (BPI)	USA	BPI is part of the International Compostability Certification Network, which includes similar trade groups in Europe, Japan, China and Taiwan
Biodegradable Plastics Society (BPS)	Japan	DIN CERTCO, BPI, BMG
Biodegradable materials Group (BMG)	China	Agreement with BPS on mutual recognition of test reports
Environmentally Biodegradable Plastics Association (EBPA)	Taiwan	Memorandum of understanding about future cooperation

(PLA), polycaprolactone (PCL), aliphatic copolyesters (poly(butylene succinate), poly(butylene succinate-*co*-adipate), poly(ethylene succinate)), aliphatic–aromatic copolyesters (poly(butylene adipate-*co*-terephthalate)), starch-based polymers, poly(vinyl alcohol), cellulose acetate.

5.5. COOPERATION BETWEEN CERTIFICATION AND LABELLING SYSTEMS

The worldwide cooperation of certification systems and the mutual recognition of certificates among institutions have been established [32]. For example, in December 2001 the Biodegradable Plastics Society (BPS) (Japan) began cooperating with DIN CERTCO, the German certification organization and the International Biodegradable Products Institute (BPI), the US certification organization, to reciprocally use testing results [31].

A cooperative network of certification institutions has been launched (Table 5.10) at a global level to facilitate trade and the application of certified products by mutual recognition of others' certificates. European Bioplastics promotes the implementation of a unified certification and labelling scheme of bioplastic products in Europe [33]. European Bioplastics (the successor to IBAW – International Biodegradable Polymers Association and Working Group, founded in 1993), registered in 2006, is the European branch association representing industrial manufacturers, processors and users of bioplastics and biodegradable polymers and its derivative products.

REFERENCES

[1]   www.epa.gov
[2]   ASTM D 6400-04 "Standard specification for compostable plastics".
[3]   ASTM D 6002-96 (Reapproved 2002)[e1] "Standard guide for assessing the compostability of environmentally degradable plastics".
[4]   ISO/DIS 17088 – Specifications for compostable plastics.
[5]   ISO 14851-1:2005 – Determination of the ultimate aerobic biodegradability of plastic materials under controlled composting conditions – Method by analysis of evolved carbon dioxide. Part 1: General method.
[6]   ISO/DIS 14851-2 – Determination of the ultimate aerobic biodegradability of plastic materials under controlled composting conditions – Method by analysis of evolved carbon dioxide. Part 2: Gravimetric measurement of carbon dioxide evolved in a laboratory-scale test.
[7]   ISO 20200:2004 – Plastics – Determination of the degree of disintegration of plastic materials under simulated composting conditions in a laboratory-scale test.
[8]   British Standards Institution's "Publicly Available Specification for Composted Materials" (PAS100); www.compost.org.uk
[9]   www.compost.org
[10]   Cooperband L.: *The Art and Science of Composting*, University of Wisconsin-Madison, Center for Integrated Agricultural Systems, 29 March 2002.
[11]   Eder P.: Case study – compost, in: *End of Waste Project*, Institute for Prospective Technological Studies, DG Joint Research Institute, September 2006, http://www.jrc.ec.europa.eu
[12]   Sharma V.K.; Canditelli M., Fortuna F., Cornacchia G.: Processing of urban and agro-industrial residues by aerobic composting: review, *Energy Conv. Mgmt.* 38 (1997) 453.
[13]   Coker C.: Composting industrial and commercial organics, 20 April 2000, www.owr.ehnr.state.nc.us/compost
[14]   Diaz L., Savage G., Chiumetti A.: Variety is spice of in-vessel life, *BioCycle* 46 (2005) 40.
[15]   Tuomela M.: Academic dissertation in microbiology, "Degradation of lignin and other ^{14}C-labelled compounds in compost and soil with an emphasis on white-rot fungi, Helsinki 2002, http://ethesis.helsinki.fi
[16]   Cheng C., Zhen L.: Controllable techniques and installations for organic-refuse composting, *Resour. Conserv.* 13 (1987) 175.
[17]   www.css.cornell.edu/compost
[18]   Richard T.L.: Municipal solid waste composting: biological processing, Cornell Waste Management Institute, www.css.cornell.edu/compost
[19]   Lasaridi K., Protopapa I., Kotsou M., Pilidis G., Manios T., Kyriacou A.: Quality assessment of compost in the Greek market: the need for standards and quality assurance, *J. Environm. Management* 80 (2006) 58.
[20]   Tognetti C., Mazzarino M.J., Laos F.: Improving quality of municipal organic waste compost, *Biores. Technol.* 98 (2007) 1067.
[21]   Assessment of options and requirements for stability and maturity testing of composts, Issue 2 – Mar 2005, published by The Waste and Resources Action Programme, www.wrap.org.uk.

[22]   Hogg D., Barth J., Favoino E., Centemero M., Caimi V., Amlinger F., Devlieger W., Brinton W., Antler S.: Comparison of compost standards within the EU, North America and Australasia, The Waste and Resources Action Programme (WRAP), June 2002, Oxon, UK.

[23]   Körner I., Braukmeier J., Herrenklage J., Leikam K., Ritzkowski M., Schlegelmilch M., Stegmann R.: Investigation and optimization of composting processes – test systems and practical examples, *Waste Management* 23 (2003) 17.

[24]   Briton W.F.: Compost quality standards & guidelines. Final Report, Woods End Research Laboratory, December 2000.

[25]   Eder P.: Case study – compost, in: *End of Waste Project*, Institute for Prospective Technological Studies, DG Joint Research Institute, September 2006, http://www.jrc.ec.europa.eu

[26]   European Parliament and Council Directive 94/62/EC of 20 December 1994 on packaging and packaging waste.

[27]   Directive 2004/12/EC of the European Parliament and Council of 11 February 2004 amending Directive 94/62/EC of 20 December 1994 on packaging and packaging waste.

[28]   Decision 2001/524/EC relating to the publication of references for standards EN 13428:2000, EN 13429:2000, EN 13430: 2000, EN 13431: 2000 and EN 13432: 2000 in the *Official Journal of the European Communities* in connection with Directive 94/62/EC in packaging and packaging waste (Official Journal L 190 of 12.07.2001).

[29]   EN 13432:2000 Packaging – Requirements for packaging recoverable through composting and biodegradation – Test scheme and evaluation criteria for the final acceptance of packaging.

[30]   DIN V 54900: Testing of compostability of plastics.

[31]   Report from the Government of Japan on the EU – Japan Business Dialogue Round Table (the BDRT) recommendations, March 2006.

[32]   Reske J.: Development of an international network for the certification of compostability. Annual Meeting of the Environmentally Biodegradable Polymer Association (EBPA), Taichung, Taiwan, December 2003.

[33]   www.european-bioplastics.org

Chapter 6
Biodegradability testing of compostable polymer materials

Chapter 6
Biodegradability testing of compostable polymer materials

6.1. DEFINITIONS RELATED TO BIODEGRADATION TESTING

Activated sludge (ISO 14851)
Biomass produced in the aerobic treatment of waste water by the growth of bacteria and other microorganisms in the presence of dissolved oxygen.

Activated vermiculite (ISO 14855-1)
Vermiculite colonized by an active microbial population during a preliminary growth phase.

Biochemical oxygen demand (BOD) (ISO 14851)
The mass concentration of the dissolved oxygen consumed under specified conditions by the aerobic biological oxidation of a chemical compound or organic matter in water, expressed as milligrams of oxygen uptake per milligram or gram of test compound.

Biodegradation phase (ISO/DIS 14855 part 2)
Time, measured in days, from the end of the lag phase of a test until about 90% of the maximum level of biodegradation has been reached.

Digested sludge (ISO 14853)
Mixture of settled sewage and activated sludge which has been incubated in an anaerobic digester at about 35°C to reduce the biomass and odour and to improve the dewaterability of the sludge. Digested sludge contains an association of anaerobic fermentation and methanogenic bacteria producing carbon dioxide and methane.

Dissolved inorganic carbon (DIC)(ISO 14852)
That part of inorganic carbon in water that cannot be removed by specific phase separation, for example by centrifugation at $40\,000\,\text{m}\cdot\text{s}^{-2}$ for 15 min or by membrane filtration using membranes with pores of $0.2\,\mu\text{m}$ to $0.45\,\mu\text{m}$ diameter.

Dissolved organic carbon (DOC) (ISO 14851)
That part of the organic carbon in water which cannot be removed by specified phase separation, for example by centrifugation at $40\,000\,\text{m}\cdot\text{s}^{-2}$ for 15 min or by membrane filtration using membranes with pores of $0.2\,\mu\text{m}$ to $0.45\,\mu\text{m}$ diameter.

Inorganic carbon (IC) (ISO 14853)
Inorganic carbon which is dissolved or dispersed in the aqueous phase of a liquid and is recoverable from the supernatant liquid after the sludge has been allowed to settle.

Lag phase (ISO/DIS 14855 part 2)
Time, measured in days, from the start of a test until adaptation and/or selection of the degradation microorganisms is achieved and the degree of biodegradation of a chemical compound or organic matter has increased to about 10% of the maximum level of biodegradation.

Maximum level of biodegradation (ISO/DIS 14855 part 2)
Degree of biodegradation, measured as a percentage, of a chemical compound or organic matter in a test, above which no further biodegradation takes place during the test.

Plateau phase (ISO/DIS 14855 part 2)
Time, measured in days, from the end of the biodegradation phase until the end of the test.

Primary anaerobic biodegradation (ISO 1485)
Structural change (transformation) of a chemical compound by microorganisms, resulting in the loss of a specific property.

Theoretical amount of evolved biogas (Thbiogas) (ISO 14853)
Maximum theoretical amount of biogas ($CH_4 + CO_2$) evolved after complete biodegradation of an organic material under anaerobic conditions, calculated from the molecular formula and expressed as millilitres of biogas evolved per milligram of test material under standard conditions.

Theoretical amount of evolved carbon dioxide (ThCO₂) (ISO/DIS 17088, ISO/DIS 14855 part 2)
Maximum theoretical amount of carbon dioxide evolved after completely oxidizing a chemical compound, calculated from the molecular formula and expressed as milligrams of carbon dioxide evolved per milligram or gram of test compound.

Theoretical amount of evolved methane (ThCH₄) (ISO 14853)
Maximum theoretical amount of methane evolved after complete reduction of an organic material, calculated from the molecular formula and expressed as milligrams of methane evolved per milligram of test material.

Theoretical oxygen demand (ThOD) (ISO 14851)
The theoretical maximum amount of oxygen required to oxidize a chemical compound completely, calculated from the molecular formula, expressed as milligrams of oxygen uptake per milligram or gram of test compound.

Total dry solids (ISO/DIS 17088; ISO/DIS 14855 part 2)
Amount of solids obtained by taking a known volume of test material or compost and drying at about 105°C to constant mass.

Total organic carbon (TOC)(ISO 14851)
All the carbon present in organic matter which is dissolved or suspended in water.

Ultimate aerobic biodegradation (ISO 14853)
Breakdown of an organic compound by microorganisms in the absence of oxygen to carbon dioxide, methane, water and mineral salts of any other elements present (mineralization) plus new biomass.

Ultimate aerobic biodegradation (ISO/DIS 17088; ISO/DIS 14855 part 2)
Breakdown of an organic compound by microorganisms in the presence of oxygen into carbon dioxide, water and mineral salts of any other elements present (mineralization) plus new biomass.

Volatile solids (ISO/DIS 17088)
Amount of solids obtained by subtracting the residue of a known volume of test material or compost after incineration at about 550°C from the total dry solids of the same sample.

6.2. INTERNATIONAL STANDARDS RELATED TO COMPOSTING

Internationally recognized standardization bodies, such as the International Organization for Standardization (ISO), as well as regional standardization bodies, such as the American Society for Testing and Materials (ASTM) and the European Committee for Standardization (CEN), are actively involved in developing standards related to composting and biodegradation. In addition, national standardization bodies, such as the German Deutsches Institut für Normung (DIN) and the Biodegradable Plastics Society (BPS) of Japan, contribute to the development and issuing of standards on compostable polymers. Recently, interest in developing national standards related to compostability, and biodegradation testing appeared in other regions of the world, e.g. in China, Taiwan and Australia.

Several ISO standards for determining the ultimate aerobic/anaerobic biodegradability of plastic materials have been published. In particular, ISO 14855-1 is a common test method that measures evolved carbon dioxide using such methods as continuous infrared analysis, gas chromatography or titration.

Table 6.1. ISO standards related to composting

Standard	Title
ISO/DIS 17088	Specifications for compostable plastics
ISO 14021:1999	Environmental labels and declarations – Self-declared environmental claims (Type II environmental labelling)
ISO 14851:1999 ISO 14851:1999/ Cor 1:2005	Determination of the ultimate aerobic biodegradability of plastic materials in an aqueous medium – Method by measuring the oxygen demand in a closed respirometer
ISO 14852:1999	Determination of the ultimate aerobic biodegradability of plastic materials in an aqueous medium – Method by analysis of evolved carbon dioxide
ISO 14853:2005	Plastics – Determination of the ultimate anaerobic biodegradation of plastic materials in an aqueous system – Method by measurement of biogas production
ISO 14855-1:2005	Determination of the ultimate aerobic biodegradability of plastic materials under controlled composting conditions – Method by analysis of evolved carbon dioxide – Part 1: General method
ISO/DIS 14855-2	Determination of the ultimate aerobic biodegradability of plastic materials under controlled composting conditions – Method by analysis of evolved carbon dioxide – Part 2: Gravimetric measurement of carbon dioxide evolved in a laboratory-scale test
ISO 15985:2004	Plastics – Determination of the ultimate anaerobic biodegradation and disintegration under high-solids anaerobic-digestion conditions – Method by analysis of released biogas
ISO 16929:2002	Determination of the degree of disintegration of plastic materials under defined composting conditions in a pilot-scale test
ISO 17556:2003	Determination of the ultimate aerobic biodegradability in soil by measuring the oxygen demand in a respirometer or the amount of carbon dioxide evolved
ISO 20200:2004	Determination of the degree of disintegration of plastic materials under simulated composting conditions in a laboratory-scale test

Table 6.2. EN standards related to biodegradation and composting

Standard	Title
EN ISO 14851:2004	Determination of the ultimate aerobic biodegradability of plastic materials in an aqueous medium – Method by measuring the oxygen demand in a closed respirometer
EN ISO 14852:2004	Determination of the ultimate aerobic biodegradability of plastic materials in an aqueous medium – Method by analysis of evolved carbon dioxide
EN ISO 14855:2004	Determination of the ultimate aerobic biodegradability and disintegration of plastic materials under controlled composting conditions – Method by analysis of evolved carbon dioxide
EN ISO 17556:2004	Determination of the ultimate aerobic biodegradability in soil by measuring the oxygen demand in a respirometer or the amount of carbon dioxide evolved
EN ISO 20200:2005	Determination of the degree of disintegration of plastic materials under simulated composting conditions in a laboratory-scale test

Table 6.3. EN standards related to packaging and composting

Standard	Title
EN 14045:2003	Packaging – Evaluation of the disintegration of packaging materials in practical oriented tests under defined composting conditions
EN 14046:2003	Packaging – Evaluation of the ultimate aerobic biodegradability of packaging materials under controlled composting conditions – Method by analysis of released carbon dioxide
EN 14806:2005	Preliminary evaluation of the disintegration of packaging materials under simulated composting conditions in a laboratory-scale test

Table 6.4. ASTM standards related to composting and biodegradation

Standard	Title
ASTM D6400-04	Standard specification for compostable plastics
ASTM D 6002-96(2002)[e1]	Standard guide for assessing the compostability of environmentally degradable plastics
ASTM D 6868-03	Standard specification for biodegradable plastics uses as coatings on paper and other compostable substrates
ASTM D 6094-97(2004)	Standard guide to assess the compostability of environmentally degradable non-woven fabrics
ASTM D 6340-98	Standard test methods for determining aerobic biodegradation of radiolabelled plastic materials in an aqueous or compost environment
ASTM D 6776-02	Standard test method for determining anaerobic biodegradability of radiolabelled plastic materials in a laboratory-scale simulated landfill environment
ASTM D 6954-04	Standard guide for exposing and testing plastics that degrade in the environment by a combination of oxidation and biodegradation
ASTM D 7081-05	Standard specification for non-floating biodegradable plastics in the marine environment

Table 6.4. (*Continued*)

Standard	Title
ASTM D 5210-92(2000)	Standard test method for determining the anaerobic biodegradation of plastic materials in the presence of municipal sewage sludge
ASTM D 5929-96(2004)	Standard test method for determining biodegradability of materials exposed to municipal solid waste composting conditions by compost respirometry
ASTM D 5338-98(2003)	Test method for determining aerobic biodegradation of plastic materials under controlled composting conditions
ASTM D 5526-94(2002)	Test method for determining anaerobic biodegradation of plastic materials under controlled landfill conditions
ASTM D 5988-03	Standard test method for determining aerobic biodegradation in soil of plastic materials or residual plastic materials after composting
ASTM D 5271-02	Standard test method for determining the aerobic biodegradation of plastic materials in an activated/sludge/wastewater/treatment system
ASTM D 6691:01	Test method for determining aerobic biodegradation of plastic in the marine environment by a defined microbial consortium
ASTM D 5511-02	Test method for determining anaerobic biodegradation of plastic materials under high-solids anaerobic-digestion conditions

6.3. PRINCIPLES OF MAIN STANDARDS RELATED TO COMPOSTING AND BIODEGRADABILITY TESTING

ISO 14855-1:2005 – Determination of the ultimate aerobic biodegradability of plastic materials under controlled composting conditions – Method by analysis of evolved carbon dioxide– Part 1: General method

Scope: This standard specifies a method for the determination of the ultimate aerobic biodegradability of plastics, based on organic compounds, under controlled composting conditions by measurement of the amount of carbon dioxide evolved and the degree of disintegration of the plastic at the end of the test. This method is designed to simulate typical aerobic composting conditions for the organic fraction of solid mixed municipal waste. The test material is exposed to an inoculum which is derived from compost. The composting takes place in an environment wherein temperature, aeration and humidity are closely monitored and controlled. The test method is designed to yield the percentage conversion of the carbon in the test material to evolved carbon dioxide as well as the rate of conversion.

It contains also a variant of the method, using a mineral bed (vermiculite) inoculated with thermophilic microorganisms obtained from compost with a specific activation phase, instead of mature compost. This variant is designed to yield the percentage of carbon in the test substance converted to carbon dioxide and the rate of conversion.

Principle: The test method determines the ultimate biodegradability and degree of disintegration of test material under conditions simulating an intensive aerobic composting process. The inoculum used consists of stabilized, mature compost derived, if possible, from composting the organic fraction of solid municipal waste.

The test material is mixed with the inoculum and introduced into a static composting vessel where it is intensively composted under optimum oxygen, temperature and moisture conditions for a test period not exceeding six months.

During the aerobic biodegradation of the test material, carbon dioxide, water, mineral salts and new microbial cellular constituents (biomass) are the ultimate biodegradation products. The carbon dioxide produced is continuously monitored, or measured at regular intervals, in test and blank vessels to determine the cumulative carbon dioxide production. The percentage biodegradation is given by the ratio of the carbon dioxide produced from the test material to the maximum theoretical amount of carbon dioxide that can be produced from the test material. The maximum theoretical amount of carbon dioxide produced is calculated from the measured total organic carbon (TOC) content. The percentage biodegradation does not include that amount of carbon converted to new cell biomass which is not metabolized in turn to carbon dioxide during the course of the test.

Additionally, the degree of disintegration of the test is determined at the end of the test, and the loss in mass of the test material may also be determined.

Vermiculite should be used instead of mature compost:

- whenever the determination of the degree of biodegradation is affected by a priming effect induced by the test material and/or
- when performing a final carbon balance with biomass determination and retrieval of the residual test material.

Priming effect: The organic matter present in large amounts in the mature compost can undergo polymer-induced degradation, known as the "priming effect", which affects the measurement of the biodegradability.

The inorganic vermiculite bed substantially reduces the priming effect, thus improving the reliability of the method. A further advantage of using vermiculite is the very small amount of carbon dioxide evolved in the blank vessel (nearly zero), because of the low level of microbial activity. This permits low levels of degradation activity to be evaluated precisely. The mineralization rates obtained with the activated vermiculite are identical, or very similar, to those obtained with mature compost, both in terms of the final degradation level and the degradation rate.

ISO/DIS 14855-2 – Determination of the ultimate aerobic biodegradability of plastic materials under controlled composting conditions – Method by analysis of evolved carbon dioxide – Part 2: Gravimetric measurement of carbon dioxide evolved in a laboratory-scale test.

In order to ensure the activity of compost inoculum, inert material which works as soil texture is mixed into compost inoculum. The carbon dioxide evolved from the test vessel is determined by using gravimetric analysis of carbon dioxide absorbent. The method, which consists of a closed system to capture evolved carbon dioxide, is available to determine the ultimate aerobic biodegradability of plastic materials under controlled composting conditions in a laboratory-scale test. The valuable information of degradation on the molecular structure of copolymers can frequently be obtained by means of isotopic labelling studies based on this test method of a closed system.

Scope: This test method specifies a method for determining the ultimate aerobic biodegradability of plastic materials in controlled composting conditions by gravimetric measurement of the amount of evolved carbon dioxide.

Principle: The method is designed to yield an optimum degree of biodegradability by adjusting the humidity, aeration ratio and temperature in a composting vessel. It also aims to determine the ultimate biodegradability of the test material by using a small-scale reactor. The degradation rate is periodically measured by increasing the weight of the evolved carbon dioxide using an absorption column charged with soda lime and soda talc on an electronic balance. The test material is mixed with the inoculum derived from mature compost and inert material such as sea sand. The sea sand takes an active part of the holding body for humidity and microorganism activity.

The amount of carbon dioxide evolved is measured at intervals on the electronic balance and the carbon dioxide content is determined. The level of biodegradation, expressed as a percentage, is determined by comparing the amount of carbon dioxide evolved with the theoretical amount ($ThCO_2$).

The test is terminated when the plateau phase of biodegradation has been attained; the standard time for termination is 45 days, but the test could continue for six months, at the latest.

ISO 20200:2004 – Plastics – Determination of the degree of disintegration of plastic materials under simulated composting conditions in a laboratory-scale test

Scope: This standard specifies a method of determining the degree of disintegration of plastic materials when exposed to a laboratory composting environment. The method is not applicable to the determination of the biodegradability of plastic materials under composting conditions.
Principle: The method determines the degree of disintegration of test materials on a laboratory scale under conditions simulating an intensive aerobic composting process. The solid matrix used consists of a synthetic solid waste inoculated with mature compost taken from a commercial composting plant. Pieces of the plastic test material are composted with this prepared solid matrix. The degree of disintegration is determined after a composting cycle, by sieving the final matrix through a 2 mm sieve in order to recover the non-disintegrated material. The reduction in mass of the test sample is considered as disintegrated material and used to calculate the degree of disintegration.

EN ISO 14851:2004 – Determination of the ultimate aerobic biodegradability of plastic materials in an aqueous medium – Method by measuring the oxygen demand in a closed respirometer (ISO 14851:1999)
ISO 14851:1999/Cor 1:2005

Scope: This Standard specifies a method by measuring the oxygen demand in a closed respirometer, for the determination of the degree of aerobic biodegradability of plastic materials, including those containing formulation additives. The test material is exposed in an aqueous medium under laboratory conditions to an inoculum from activated sludge, compost or soil.
If an unadapted sludge is used as the inoculum, the test simulates the biodegradation processes which occur in a natural aqueous environment; if a mixed or pre-exposed inoculum is used, the method can be used to investigate the potential biodegradability of a test material.
Principle: The biodegradability of a plastic material is determined using aerobic microorganisms in an aqueous system. The test mixture contains an inorganic medium, the organic test material (the sole source of carbon and energy) with a concentration between 100 mg/l and 2000 mg/l of organic carbon, and activated sludge or a suspension of active soil or compost as the inoculum. The mixture is stirred in closed flasks in a respirometer for a period not exceeding six months. The carbon dioxide evolved is absorbed in a suitable absorber in the headspace of the flasks. The consumption of oxygen (BOD) is determined, for example by measuring the amount of oxygen required to maintain a constant volume of gas in the respirometry flasks, or by measuring the change in volume or pressure (or a combination of the two) either automatically or manually.

The level of biodegradation is determined by comparing the BOD with the theoretical amount (ThOD) and expressed in per cent. The influence of possible nitrification processes on the BOD has to be considered. The test result is the maximum level of biodegradation determined from the plateau phase of the biodegradation curve. There is the possibility of improving the evaluation of biodegradability by calculating a carbon balance.

ISO 14852:1999 – Determination of the ultimate aerobic biodegradability of plastic materials in an aqueous medium – Method by analysis of evolved carbon dioxide

Scope: This Standard specifies a method, by measuring the amount of carbon dioxide evolved, for the determination of the degree of aerobic biodegradability of plastic materials, including those containing formulation additives. The test material is exposed in a synthetic medium under laboratory conditions to an inoculum from activated sludge, compost or soil. If an unadapted activated sludge is used as the inoculum, the test simulates the biodegradation processes which occur in a natural aqueous environment; if a mixed or pre-exposed inoculum is used, the method can be used to investigate the potential biodegradability of a test material. The standard is designed to determine the potential biodegradability of plastic materials or give an indication of their biodegradability in natural environments.

The method enables the assessment of the biodegradability to be improved by calculating a carbon balance.

Principle: The biodegradability of a plastic material is determined using aerobic microorganisms in an aqueous system. The test mixture contains an inorganic medium, the organic test material (the sole source of carbon and energy) with a concentration between 100 mg/l and 2000 mg/l of organic carbon, and activated sludge or a suspension of active soil or compost as the inoculum. The mixture is agitated in test flasks and aerated with carbon dioxide-free air over a period of time depending on the biodegradation kinetics, but not exceeding six months. The carbon dioxide evolved during the microbial degradation is determined by a suitable analytical method. For example, the carbon dioxide evolved is absorbed in sodium hydroxide (NaOH) solution and determined as dissolved inorganic carbon (DIC) using, e.g., a DOC analysed without incineration. Another use is the titrimetric method using a barium hydroxide solution.

The level of biodegradation is determined by comparing the amount of carbon dioxide evolved with the theoretical amount (ThCO$_2$) and expressed in per cent. The test result is the maximum level of biodegradation, determined from the plateau phase of the biodegradation curve. Optionally, a carbon balance may be calculated to give additional information on the biodegradation.

The Standard is specially designed for the determination of the biodegradability of plastic materials. There is a possibility of improving the evaluation of the biodegradability by calculating a carbon balance.

ISO 14853:2005 – Plastics – determination of the ultimate anaerobic biodegradation of plastic materials in an aqueous system – Method by measurement of biogas production

Scope: This Standard specifies a method for the determination of the ultimate anaerobic biodegradability of plastics by anaerobic microorganisms. The test calls for exposure of the test material to sludge for a period of up to 60 days, which is longer than the normal sludge retention time (25 to 30 days) in anaerobic digesters, though digesters at industrial sites can have much longer retention times.

Principle: The biodegradability of a plastic material is determined using anaerobic conditions in an aqueous system. Test material with a concentration of 20 mg/l to 200 mg/l organic carbon (OC) is incubated at 35°C ± 2°C in sealed vessels together with digested sludge for a period normally not exceeding 60 days. Before use, the digested sludge is washed so that it contains very low amounts of inorganic carbon (IC) and diluted to 1 g/l to 3 g/l total solids concentration. The increase in headspace pressure or the volumetric increase (depending on the method used for measuring biogas evolution) in the test vessels resulting from the production of carbon dioxide and methane is measured. A considerable amount of carbon dioxide will be dissolved in water or transformed to bicarbonate or carbonate under the conditions of the test. The inorganic carbon (IC) is measured at the end of the test. The amount of microbiologically produced biogas carbon is calculated from the net biogas production and the net IC formation in excess of blank values. The percentage biodegradation is calculated from the total amount of carbon transformed to biogas and IC and the measured or calculated amount added as test material. The course of biodegradation can be followed by making intermediate measurements of biogas production. As additional information, the primary biodegradability can be determined by specific analyses at the beginning and end of the test.

The test method is designed to determine the biodegradability of plastic materials under anaerobic conditions. Optionally, the assessment of the recovery rate may also be determined.

Reference material: Anaerobically biodegradable polymer, e.g. poly-β-hydoroxybutyrate, cellulose or poly(ethylene glycol) 400.

ISO 15985:2004 Plastics – Determination of the ultimate anaerobic biodegradation and disintegration under high-solids anaerobic-digestion conditions – Method by analysis of released biogas

Scope: This Standard specifies a method for the evaluation of the ultimate anaerobic biodegradability of plastics based on organic compounds under high-solids anaerobic-digestion conditions by measurement of evolved biogas and the degree of disintegration at the end of the test. This method is designed to simulate typical anaerobic digestion conditions for the organic fraction of mixed municipal solid waste. The test material is exposed in a laboratory test to a methanogenic inoculum derived from anaerobic digesters operating only on pretreated household waste. The anaerobic decomposition takes place under high-solids (more than 20% total solids) and static non-mixed conditions. The test method is designed to yield the percentage of carbon in the test material and its rate of conversion to evolved carbon dioxide and methane (biogas).

Principle: The test method is designed to be an optimized simulation of an intensive anaerobic digestion process and determines the ultimate biodegradability and degree of disintegration of a test material under high-solids anaerobic conditions. The methanogenic inoculum is derived from anaerobic digesters operating on pretreated household waste, preferably only the organic fraction.

The test material is mixed with the inoculum and introduced into a static digestion vessel where it is intensively digested under optimum temperature and moisture conditions for a test period of 15 days or longer until a plateau in net biodegradation has been reached.

During the anaerobic biodegradation of the test material, methane, carbon dioxide, water, mineral salts and new microbial cellular constituents (biomass) are produced as the ultimate biodegradation products. The biogas (methane and carbon dioxide) evolved is continuously monitored or measured at regular intervals in test and blank vessels to determine the cumulative biogas production. The percentage biodegradation is given by the ratio of the amount of biogas evolved from the test material to the maximum theoretical amount of biogas that can be

Table 6.5. Summary of biodegradability and composting methods

Standard	Medium	Duration	Temperature	Reference material	Measurements
ISO 14855-1:2005	Mature compost, optionally vermiculite	Not exceeding six months	58 ± 2°C	Thin-layer chromatography grade cellulose as positive reference	1. CO_2 evolution (by IR analysis, gas chromatography, titration method, etc.) 2. disintegration (visual evaluation, relevant physical properties measurements)
ISO/DIS 14855-2	Mature compost + inert material (sea sand)	Standard time (45 days); up to six months is allowed	58 ± 2°C	Thin-layer chromatography grade cellulose as positive reference	CO_2 evolution (by gravimetric method)
EN ISO 14851: 2004	Aqueous	Not exceeding six months	Preferably between 20 and 25°C	Aniline, microcrystalline cellulose powder, ashless cellulose filters or poly-β-hydroxybutyrate as positive reference	Oxygen consumption (by, for example, respirometric method or measurements of changes in volume or pressure)
EN ISO 14852-1999	Aqueous	Not exceeding six months	Preferably between 20 and 25°C	Aniline, microcrystalline cellulose powder, ashless cellulose filters or poly-β-hydroxybutyrate as positive reference	CO_2 evolution (CO_2 or DIC analyser or apparatus for titrimetric determination after complete absorption in a basic solution)

produced from the test material. The maximum theoretical amount of biogas produced is calculated from the measured total organic carbon (TOC). This percentage biodegradation does not include the amount of carbon converted to new cell biomass which is not metabolized in turn to biogas during the course of the test.

Additionally, the degree of disintegration of the test material is determined at the end of the test and the loss in mass of the test material may also be determined.

Reference material: Thin-layer chromatography grade cellulose with a particle size of less than $20\,\mu m$ as the positive reference material.

ISO 17556:2003 – Determination of the ultimate aerobic biodegradability in soil by measuring the oxygen demand in a respirometer or the amount of carbon dioxide evolved

Scope: This Standard specifies a method for determining the ultimate aerobic biodegradability of plastic materials in soil by measuring the oxygen demand in a closed respirometer or the amount of carbon dioxide evolved. This method is designed to yield an optimum degree of biodegradation by adjusting the humidity of the test soil.

If a non-adapted soil is used as an inoculum, the test simulates the biodegradation processes which take place in a natural soil environment; if a pre-exposed soil is used, the method can be used to investigate the potential biodegradability of a test material.

Principle: This method is designed to yield the optimum rate of biodegradation of a plastic material in a test soil by controlling the humidity of the soil, and to determine the ultimate biodegradability of the test material.

The plastic material, which is the sole source of carbon and energy, is mixed with the soil. The mixture is allowed to stand in a flask over a period of time during which the amount of oxygen consumed (BOD) or the amount of carbon dioxide evolved is determined. The BOD is determined, for example, by measuring the amount of oxygen required to maintain a constant gas volume in a respirometer flask, or by measuring either automatically or manually the change in volume or pressure (or a combination of the two). The amount of carbon dioxide evolved is measured at intervals dependent on the biodegradation kinetics of the test substance by passing carbon dioxide-free air over the soil and then determining the carbon dioxide content of the air by a suitable method.

The level of biodegradation, expressed in per cent, is determined by comparing the BOD with the theoretical oxygen demand (ThOD) or by comparing the amount of carbon dioxide evolved with the theoretical amount ($ThCO_2$). The influence of possible nitrification processes on the BOD has to be considered. The test is terminated when a constant level of biodegradation has been attained or, at the latest, after six months.

6.4. COMPOSTING AT LABORATORY SCALE

The composting test method based on activated vermiculite was proposed as a comprehensive system for the assessment of the environmental impact of compostable polymers [1, 2]. Vermiculite, a clay mineral, can be activated (by an inoculation with an appropriate microbial population and fermentation) and used as a solid matrix in place of mature compost in the controlled composting test. The formula of vermiculite is: $(Mg,Fe,Al)_3(Al,Si)_4O_{10}(OH)_2.4H_2O$. The results obtained with two materials (cellulose and a starch-based blend) indicated that activated vermiculite affected neither the biodegradation rate nor the final biodegradation level.

On the other hand, possible metabolic intermediates and polymeric residues left after biodegradation could be recovered more easily from activated vermiculite than from mature compost, a very complex organic matter. Therefore, at test termination it was possible to determine the carbon balance taking into account both the evolved CO_2 and a polymeric residue extracted from vermiculite, totalling 101% of the carbon present originally in the test material. To conclude, it allows, in a single test, (i) the measurement of the mineralization of the polymer under study; (ii) the retrieval of the final polymeric residues; (iii) determination of the biomass (to make a final mass balance); and (iv) detection of breakdown products of the original polymer. The vermiculite test method is also suitable to perform ecotoxicological studies [2].

Different vermiculite media were studied in order to determine the parameters of an inert solid medium which could simulate the degradation of a polymer in compost [3]. Five different vermiculite media have been tested according to type of activation and the amount of inoculum used. The mineralization curves obtained for simulation tests have been compared with the mineralization curve of starch biodegradation in compost.

Glucose, starch, and cellulose can increase the biodegradation of the compost used as a solid matrix in the biodegradation test under composting conditions (priming effect). The enhanced evolution of carbon dioxide determines an overestimation of the biodegradation of the starch- and cellulose-based materials and, in some cases, values higher than 100% can be reached. Therefore, it was verified that by using activated vermiculite, an inorganic matrix, the priming effect can be reduced, improving the reliability of the test method [4]. Glucose, the most effective primer, causes the attainment of biodegradation values significantly higher than 100% in mature compost while this does not happen in activated vermiculite. Since all the initial carbon present in the activated vermiculite was converted into CO_2 within the test period, it was concluded that a substantial priming effect cannot occur for the lack of organic carbon. Furthermore, by measuring in parallel both the consumption of glucose and the CO_2 evolution, the yield of CO_2 production ($Y_{CO2} = C_{CO2}/C_{glucose}$) was determined. In no case was a value higher than 1 found, a clear indication of the priming effect.

Variation of microbial population in the compost was examined at different stages of the composting [5]. Moisture content was controlled in the range $64 \pm 4\%$, and the thermophilic stage lasted about two weeks. The temperature during the composting was controlled not to exceed 58°C. In the initial stage of the composting, mesophilic strains were more numerous than thermophilic ones. As the thermophilic stage set in, thermophilic bacteria and actinomycetes outnumbered mesophilic correspondents while fungi were not detected at all. In the cooling and maturing phases, a substantial number of actinomycetes were still found. However, bacteria decreased significantly in number, and only a small number of mesophilic fungi reappeared. When glucose was added to the compost, the so-called "priming effect" was observed, in that the amount of CO_2 evolved was larger than that predicted by assuming that all added glucose was mineralized into CO_2. However, the priming effect decreased as the quantity of the glucose in the compost increased. Addition of 5 wt% of glucose to the compost increased the number of microorganisms by 10–100 times.

Specimens in film shape as well as in powder shape were subjected to the biodegradation tests to investigate dependence of the test results on the shape of the specimens [6]. Biodegradation of plastics was tested in compost made with animal fodder. Polypropylene (PP) was chosen as a non-degradable plastic. Poly(L-lactic acid) (PLLA) and poly(butylene succinate) (PBS) were selected as slowly degrading plastics while polycaprolactone (PCL) and poly(butylene succinate-*co*-adipate) were chosen as easily degradable plastics. Biodegradability of PP in film shape as well as in powder shape was tested to investigate the possible change in the microbial aspiration, because the shape of the specimens may affect aeration behaviour in the compost. Biodegradation results of PLLA and PBS depended on their shape all through the biodegradation test. In contrast, the shape of PCL and PBS exerted influences on their biodegradability only at the early stage of the biodegradation, while at the late stage, the biodegradation proceeded almost independently of their shape.

Some laboratory composting facilities were developed and described [7–9]. An automated multi-unit composting facility for studying the biodegradation of polymers was developed in accordance with the guidelines included in standards ISO/DIS 14855 and ASTM D 5338-92 [7]. In the system, cellulose, newspaper and two starch-based polymers were treated with compost in a series of $3\,dm^3$ vessels at 52°C and under conditions of optimum moisture and pH. The degradation was followed over time by measuring carbon dioxide evolved. Results showed that at 52°C over 45 days cellulose and starch-based blends degraded by 90, 87 and 72%, respectively. The cellulose and lignin-hemicellulose-based newspaper was degraded by approximately 50% under experimental conditions. A Biological Oxygen Demand (BOD) measurement system was adapted to monitor biodegradation process in solid media [8]. BOD is widely used for the examination of sewage water, effluents, polluted water and for the assessment of biodegradation of chemicals and biodegradable polymers, but exclusively in aquatic media.

After the optimization of sample concentration and test temperature, the measurement set-up possessing relatively small reaction vessels of 250 ml with 80 g of soil mix proved to supply reliable and reproducible results. The system was optimized with microcrystalline cellulose – used as reference material in aquatic and solid test as well – showing 89.3 $\pm$ 3.2% degree of degradation after 21 days. Two test systems for composting studies of different scales (up to 1500 ml; up to 100 l) were described [9]. The laboratory scale composting unit allows for the simulation of a composting process with all operating controls (aeration, moistening, turning) common to those in a composting facility. The developed set-up should simulate processes such as pressure-forced windrow and pile composting as well as tunnel, box, container, and channel systems.

The example of laboratory composting system and vessel is given in Figs 6.1 and 6.2, respectively [10]. The composting vessels were placed in the laboratory composting system. Humidified air was passed through flow meters and then into the composting vessel. External heat was applied to maintain a constant temperature of 52°C. The exhaust air was directed through a two-way valve attached to a gas chromatograph to measure CO_2 concentration. Once per week, the compost in the vessels was stirred and compost samples removed to determine the moisture content, which ranged from 48 to 55% (calculated on wet weight basis).

The medium closest to the natural condition is a solid medium (soil, compost, inert solid media) [11]. The studies on solid-state biodegradation processes in field and laboratory conditions, and in various media such as compost, soil, or inert material, were reviewed [11]. The external parameters that influence biodegradation kinetics – the material concentration in the solid medium, the environmental conditions (temperature, pH, moisture, oxygen availability, composition, and concentration of inorganic nutrients of the solid medium), the microbial population (concentration, nature, and interactions), the presence or the absence of other degradable substances, and the conditions and properties of the test system (volume and shape of the vessels) – were presented. The most significant parameters would appear to be the substrate type, moisture content, and temperature.

Maximum temperature during the thermophilic phase and moisture content were controlled in the course of composting to examine the effects of these composting conditions on the quality of the compost used for the evaluation of the biodegradability of plastics [12]. The moisture content during composting was controlled at 65%, while keeping the maximum temperature

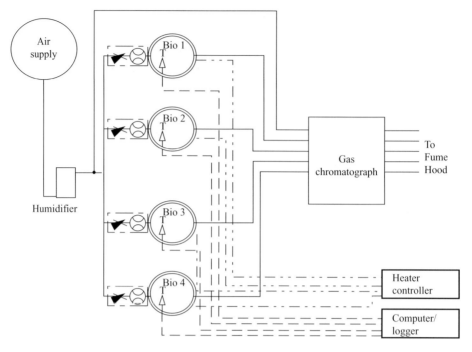

Figure 6-1 The laboratory composting system (reprinted with permission [10]).

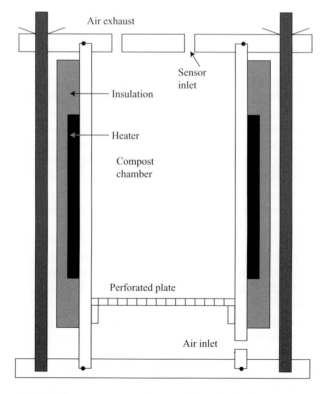

Figure 6-2 The laboratory composting vessel (reprinted with permission [10]).

below 46ºC, 58ºC and 70ºC, respectively. In turn, the maximum temperature was controlled to be below 58ºC, while maintaining the moisture content at 45, 55 and 65%, respectively. Biodegradability tests for cellulose, polycaprolactone and poly(butylene succinate-*co*-butylene adipate) were performed in the five compost samples. All three samples were biodegraded faster in the compost prepared with a maximum temperature of 45ºC than in the composts prepared at 58ºC or 70ºC, due to a larger number of microbial cells in the former compost sample. The biodegradation proceeded faster in the compost prepared with a moisture content of 65% than in the compost prepared with a moisture content of 45 and 55%.

6.5. BIODEGRADABILITY TESTING METHODS

An overview of the testing methods which have been used to evaluate biodegradability of polymers and packaging materials was given by Itävaara *et al.* [13]. Two kinds of tests for biodegradability of polymers were proposed: screening tests and tests that simulate *in situ* conditions. Screening tests include enzymatic and aquatic test under anaerobic and aerobic (Sturm test) conditions. Real-life tests are based on three compost tests (compost environment, standard compost test, and CO_2 compost test elaborated at VTT). During the first test, compostability of the materials is determined as the weight loss of the sample. Evaluation of the compostability of the samples is performed visually at weekly intervals in connection with turning the biowaste, and weight loss is measured at the end of the test when the positive control sample has been completely degraded and the temperature decreased to the outdoor temperature. The other two tests are based on CO_2 evolution.

Different polymers (e.g. polyhydroxybutyrate-hydroxyvalerate, polycaprolactone, cellulose acetate) representing varied biodegradability levels were studied using an aerobic respirometric test in order to model degradation kinetics in a liquid medium [14]. The mathematical model was proposed that fitted as well as possible the CO_2 evolution curves. Three kinetic parameters were determined: one represents the maximal percentage of carbon converted into CO_2, the second the "half-life time" in days of the degrading part of the material and the third one the curve radius.

Results of an international ring test of two laboratory methods were presented for investigating the biodegradability of organic polymeric test materials in aquatic test systems based on respirometry and the evolution of carbon dioxide [15]. These methods were developed further from the well-known standardized biodegradation tests ISO 9408 (1999) and ISO 9439 (1999). A ring test was run using a poly(caprolactone)–starch blend and an aliphatic–aromatic copolyester as test materials and a microcrystalline cellulose powder as a reference material. The most important improvements were the extension of the test period up to six months, the increase of the buffer capacity and nutrient supply of the inorganic medium, an optimization of the inoculum, and, optionally, the possibility of a carbon balance. The test methods have been meanwhile established as standards ISO 14851 (1999) and ISO 14852 (1999).

Test methods currently available for testing polymer degradability have been reviewed by Gu *et al.* [16]. Table 6.6 presents a comparison of several methods available for testing degradability of different polymers and under a range of environmental and simulation techniques [16]. The gravimetric method is the most widely used technique with a long history of success. Requirements for the polymeric materials include that the polymer should be easily moulded into some physical intact forms in sheet or strips and the specimens should not be sensitive to moisture to lose weight or easily hydrolysed significantly upon exposure in a short period of time. Since the goal of this method is to obtain gravimetric information of exposed samples, specimens taken at different time intervals may also be used for chemical characterization including molecular weight and UV-visible spectra. When additional samples can be included initially, microbiological investigation including isolation of microorganisms from surfaces, characterization of the microorganisms, molecular analysis of pure species, mixed culture or the community, can all be accomplished. The major advantage of this method is the simplicity and wide adaptability, while the drawback is that a large number of polymer samples are needed initially to carry out this kind of test.

The respirometric method measures either CO_2 produced or CO_2 consumed or both of them in an enclosed system with proper maintenance or regulation of air or oxygen supply. This technique is especially suitable for confirmation on the extent of mineralization. It can be used for measuring degradation of soluble powder from fragile polymeric materials. This method is easily adapted to a whole range of environmental conditions and/or specified or mixed culture microorganisms.

Examples of laboratory systems developed for biodegradation studies based on CO_2 evolution according to ISO 148551 and ISO/DIS 14855-2 standards are given in Figs 6.3 and 6.4, respectively [17]. The experimental set-up for biodegradation tests based on ISO 14855-1 shown in Fig. 6.3 is managed by Mitsui Chemical Analysis and Consulting Service, Inc., one of the research institutes that can determine the biodegradability of plastic products authorized by BPS for the GreenPla certification system in Japan. The CO_2 produced from the reaction vessels is trapped in alkaline solution bottles. The amounts of trapped CO_2 are determined by the titration of the acid solution to trap solutions.

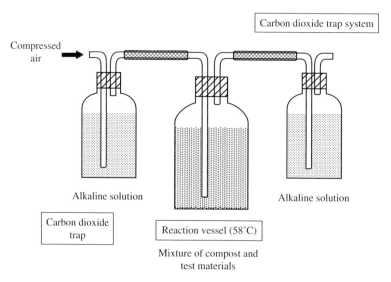

Figure 6-3 Biodegradation evaluation method based on ISO 14855-1 (reprinted with permission [17]).

The biodegradation test system with gravimetric measurement using the Microbial Oxidative Degradation Analyser (MODA) based on ISO/DIS 14855-2 uses the CO_2 trap system with CO_2 absorption column (Fig. 6.4). At first, room air is purged into a carbon dioxide trap to remove CO_2 in the air. Then, the air is moisturized and purged into the reaction vessel controlled at 58°C and 70°C using a thermosensor and ribbon heater. The air with the evolved CO_2 from biodegradation of the samples is poured into the ammonia trap to remove the produced ammonia from the compost for obtaining an accurate carbon balance using a gravimetric measurement.

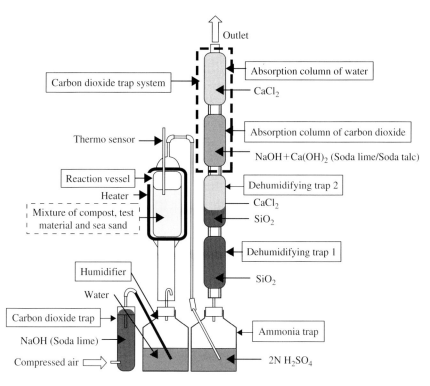

Figure 6-4 Biodegradation evaluation method by gravimetric measurement of carbon dioxide evolved in laboratory-scale test using the Microbial Oxidative Degradation Analyser (MODA) instrument in controlled compost based on ISO/DIS 14855-2 (reprinted with permission [17]).

The air with its CO_2 is poured into dehumidifying traps to remove the moisture from the stream in air for an accurate carbon weight balance and then poured into an absorption column of carbon dioxide and an absorption column of water. In these two columns with soda lime (NaOH immobilized to slaked lime) and soda talc (NaOH immobilized to talc), the produced CO_2 is absorbed by the reactions indicated in Eq. (1):

$$CO_2 + 2NaOH \rightarrow Na_2CO_3 + H_2O \qquad (1)$$

The produced H_2O is simultaneously trapped in these two columns. The weight of these two columns is increased the same as the weight of the produced CO_2, thus the produced CO_2 is easily obtained by a gravimetric method.

The enzymatic approach, based on the monitoring of pH changes in the degradation system and an increase of acidity is a strong indication of surface hydrolysis of polymers after exposure to enzyme [16]. Because this kind of system may not applicable for simulated environmental conditions involving microorganisms and the limitations of certain polymer chemistry, this method has a limited opportunity for wider applications. The advantage of this method is that a small quantity of material would be needed, especially for material in the development stage.

Electrochemical impedance spectroscopy (EIS) has been tested for monitoring biodeterioration of high strength materials and has very high sensitivity.

Table 6.6. Comparison of testing methods available for biodegradability studies of polymers [16]

Methods	Polymer forms	Inoculum and degradation criteria monitored	Comments
Gravimetry	Film or physical intact forms	A wide range of inocula can be used from soil, waters, sewage or pure species of microorganisms from culture collections	This method is robust and also good for isolation of degradative microorganisms from environment of interest. Reproducibility is high. Disintegration of polymer cannot be differentiated from biodegradation
Respirometry	Film, powder, liquid and virtually all forms and shapes	Either oxygen consumed or CO_2 produced under aerobic conditions. Under methanogenic conditions, produced methane can be monitored	This method is most adaptable to a wide range of materials. It may require a specialized instrument. When fermentation is the major mechanism of degradation, this method gives underestimation of the results
Surface hydrolysis	Films or others	Generally aerobic conditions, pure enzymes are used. Hydrogen ions (pH) released are monitored as incubation progresses	Prior information about the degradation of the polymer by microorganisms or particular enzymes is needed for the target specific test
Electochemical impedance spectroscopy	Films or coatings resistant to water	The test polymers should be adhered to surface of conductive materials and electrochemical conductance recorded	Polymer must be initially water impermeable for signal transduction. Degradation can proceed quickly and as soon as degradation is registered no further degradation processes can be distinguished

6.6. BIODEGRADATION OF BIODEGRADABLE POLYMERS FROM RENEWABLE RESOURCES

6.6.1 Biodegradation of poly(lactic acid) – PLA

Degradation mechanisms

Biodegradation of PLA proceeds via a two-stage mechanism [18]. In the first step, hydrolysis of ester linkage occurs. This step can be accelerated by acid or bases and is affected by both temperature and moisture levels [19]. In the primary degradation phase, no microorganisms are involved. As the average molecular weight diminishes, microorganisms present in the soil begin to digest the lower molecular weight lactic acid oligomers, producing carbon dioxide and water. This two-stage mechanism of degradation is a distinct advantage of PLA over other biodegradable polymers, which typically degrade by a single-step process involving bacterial attack on the polymer itself. This is a useful attribute, particularly for product storage and in applications requiring food contact. PLA degrades rapidly in the composting atmosphere of high humidity and temperature (55–70°C). But, at lower temperatures and/or lower humidity, the storage stability of PLA products is considered to be acceptable.

Degradation in compost

Polylactic acid (PLA) is fully biodegradable when composted in a large-scale operation with temperatures of 60°C and above. The first stage of degradation of PLA (two weeks) is via hydrolysis to water soluble compounds and lactic acid, then metabolization by microorganisms into carbon dioxide, water and biomass proceeds [20].

PLA is largely resistant to attack by microorganisms in soil or sewage under ambient conditions. The polymer must first be hydrolysed at elevated temperatures (>58°C) to reduce the molecular weight before biodegradation can commence. Thus, PLA will not degrade in typical garden compost. Under typical use and storage conditions PLA is quite stable [21].

The degradation of polylactic acid (PLA) plastic films in Costa Rica soil and in a leaf composting environment was investigated [22]. The average soil temperature and moisture content in Costa Rica were 27°C and 80%, respectively. The average degradation rate of PLA plastic films in the soil of the banana field was 7657 M_w/week. PLA films required two weeks to disintegrate physically in leaf compost rows.

Poly(lactide) (PLA) bottles were used as the test material to determine polymer biodegradation under simulated conditions using an automatic laboratory-scale respirometric system [23]. The results were compared with those for corn starch powder and poly(ethylene terephthalate) bottles. At 63 days of exposure at 58°C and 55% relative humidity, PLA, corn starch, and PET achieved 64.2, 72.4 and 2.7% mineralization respectively. It was stated that, based on ASTM D 6400 and ISO 14855, PLA bottles qualified as biodegradable since mineralization was greater than 60%.

The biodegradability of lactic acid-based polymers was studied under controlled composting conditions (according to future CEN EN 14046), and the quality of the compost was evaluated [24]. All the polymers biodegraded to over 90% of the positive control in six months, which is the limit set by the CEN standard.

The biodegradation of polylactide (PLLA) was studied at different elevated temperatures in aerobic and anaerobic, aquatic and solid-state conditions. In the aerobic aquatic headspace test the mineralization of PLLA was very slow at room temperature, but faster under thermophilic conditions [25]. The clear effect of temperature on the biodegradability of PLLA in the aquatic test indicates that its polymer structure has to be hydrolysed before microorganisms can utilize it as a nutrient source. At similar elevated temperatures, the biodegradation of PLLA was much faster in anaerobic solid-state conditions than in aerobic aquatic conditions. The behaviour of PLLA in the natural composting process was similar to that in the aquatic biodegradation tests, biodegradation starting only after the beginning of the thermophilic phase. These results indicate that PLLA can be considered as a compostable material, being stable during use at mesophilic temperatures, but degrading rapidly during waste disposal in compost or anaerobic treatment facilities.

It was demonstrated that PLA can be efficiently composted when added in small amounts (<30% by weight) to pre-composted yard waste (i.e. grass, wood mulch, and tree leaves in equal parts by weight) [10]. Garden waste and extruded PLA sheets were placed in laboratory composting vessels for four weeks. Evolved carbon dioxide concentration was measured by using gas chromatography to assess polymer degradation.

In all cases (0, 10, or 30% PLA), the amount of evolved CO_2 significantly increased as composting time increased (Fig. 6.5). Compost pH dropped (from 6.0 to 4.0) after four weeks of composting for 30% PLA, but remained unchanged (6.30 for 0 or 10% PLA). Most likely, in the case of 30% PLA, substantial chemical hydrolysis and lactic acid generation lowered the compost pH. The lowered pH likely suppressed microbial activity, thus explaining the lack of difference in carbon dioxide emissions between 10 and 30% PLA mixtures. The reduction in PLA molecular weight was observed after four weeks of composting (Fig. 6.6).

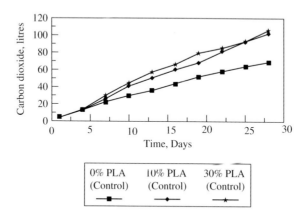

Figure 6-5 Generation of CO_2 during composting of yard waste compost/PLA mixtures (100%/0%, 90%/10%, or 70%/30% on dry weight basis). Reprinted with permission from [10].

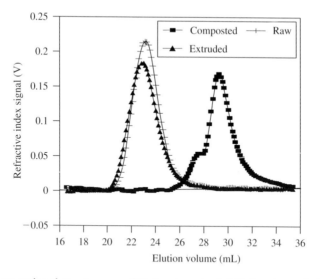

Figure 6-6 Gel permeation chromatograms of PLA resin, extruded PLA and extruded PLA composted for four weeks. Reprinted with permission from [10].

Recently, poly(lactic acid) powders were proposed as the reference test materials for the international standard of biodegradation evaluation methods [17]. Mechanical crushing at low temperature of polymer pellets using dry ice was applied as the method for producing polymer powder of PLA. After sieving the average diameter of the PLA particles was 214.2 μm. The biodegradation speeds of these PLA polymer powders were evaluated by two methods based on the international standard and one *in vitro* method based on the enzymatic degradation. First, the degree of biodegradation for the PLA powder was 91% for 35 days in a controlled compost determined by a method based on ISO 14855-1 (JIS K6953) at 58°C. Second, the polymer powders were measured for biodegradation by the Microbial Oxidative Degradation Analyser (MODA) in a controlled compost at 58°C and 70°C based on ISO/DIS 14855-2 under many conditions. The degree of biodegradation for PLA powder was approximately 80% for 50 days (Fig. 6.7).

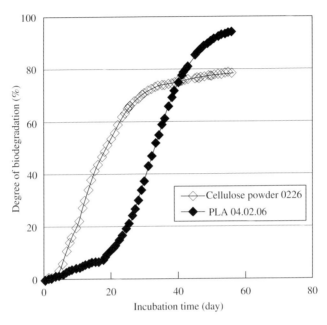

Figure 6-7 Biodegradation test of PLA and cellulose powders by ISO 14855-2 method using MODA instrument in controlled compost at 58°C. Reprinted with permission from [17].

The degradation of two commercially available biodegradable packages made of PLA was investigated and compared under real compost conditions and under ambient exposure, using visual inspection, gel permeation chromatography, differential scanning calorimetry and thermogravimetry analysis [26]. PLA bottles made of 96% L-lactide exhibited lower degradation than PLA delicatessen containers made of 94% L-lactide, mainly due to their highly ordered structure and therefore their higher crystallinity. Temperature, relative humidity and pH of the compost pile played an important role in the rate of degradation of the packages. PLA deli containers degraded in <30 days under composting conditions (temperature >60°C, RH>65%, pH ≈ 7.5).

Degradation in other environments
Polylactic acid (PLA) undergoes enzymatic or non-enzymatic hydrolysis when it is exposed to an aqueous environment. Several factors, such as temperature, pH, additives, copolymerization, initial molar mass, specimen size, residual monomer and degree of crystallinity have been reported to affect the rate of hydrolysis of PLA. The biotic and abiotic degradation of poly(L-lactide) has been studied with pyrolysis gas chromatography mass spectrometry (Py-GC/MS) [27]. It was shown that degradation in the biotic medium proceeded mainly via a surface erosion mechanism, whereas bulk erosion was the predominant degradation mechanism in the abiotic medium. Based on the SEC and PY-GC/MS data, it was reported that degradation was faster in the biotic than in the abiotic sample.

Table 6.7. Composting studies of PLA polymers

Polymer	Material description	Method used	Conditions/Results	Remarks	References
PLLA	Poly-L-lactide; Neste Oy In the form of non-woven fabrics and blown film	Bench-scale composting; carbon dioxide measurements	After 60 days final mineralization of PLLA films: 99%; PLLA fabrics: 73 and 48%	Newspaper as reference substance	[25]
PLA	PLA bottle, Biota	Composting, ISO 14855 ASTM D 6400	At 63 days of exposure at 58°C and 55% relative humidity: 64.2 % mineralization	Corn starch as positive reference	[23]
PLA	PLA film	Composting; leaf compost rows, measurement of M_w	Temperature:55–60°C; humidity: 50–70%; PLA films required two weeks to disintegrate physically in the compost rows; degradation rate 109 173 and 68–532 M_w/week		[22]
PLA	PLLA (poly(L-lactide)) – laboratory synthesized	Controlled composting test (prEN14046); CO_2 evolution measurement	Biodegradation: 92% (±17%) for PLLA in 202 days (56% in 150 days)	Whatman Chromatography paper as positive control	[24]
PLA	PLA (commercial; extruded 1.5 mm thickness sheets)	Composting; yard waste compost; CO_2 evolution measurement and molecular weight changes by GPC	Notable decrease in PLA molecular weight		[10]
PLA	Poly(lactic acid); commercial sample from Mitsui Chemicals	Composting (ISO 14855-1, ISO 14855-2, enzymatic degradation); CO_2 evolution measurement based on titration and gravimetric methods	Biodegradation of PLA powder was 91% for 31 days (ISO 14855-1 method) and 80% for 50 days at 58°C (ISO 14855-2 method)	Cellulose powder was used as reference material; PLA in the form of powders of different size was used	[17]
PLA	Poly(lactic acid); commercial bottles and deli containers	Composting under real conditions (compost pile; temp. 65°C; moisture 63%, pH 8.5); visual inspection; molecular weight changes (GPC method); glass transition and melting temperature (DSC method); decomposition temperature (TGA method).	Degradation of PLA containers <30 days under composting conditions		[26]

Polyester-degrading ability of actinomycetes obtained from culture collections was investigated by the formation of clear zones on polyester-emulsified agar plates [28]. Using 41 genera (43 strains) of actinomycetes with phylogenetic affiliations, poly(L-lacticde)-degraders were found to be limited to members of family *Pseudonocardiaceae* and related genera. On the other hand, poly(β-hydroxybutyrate)-, poly(caprolactone)-, and poly(butylene succinate)-degraders were widely distributed in many families.

Microbial and enzymatic degradation of PLA was reviewed by Tokiwa [29]. Most of the PLA-degrading microorganisms phylogenetically belong to the family of *Pseudonocardiaceae* and related genera such as *Amycolatopsis, Lentzea, Kibdelosporangium, Streptoalloteichus,* and *Saccharothrix*. Several proteinous materials such as silk fibroin, elastin, gelatin, and some peptides and amino acids were found to stimulate the production of enzymes from PLA-degrading microorganisms. In addition to proteinase K from *Tritirachium album*, subtilisin, a microbial serine protease and some mammalian serine proteases such as α-chymotrypsin, trypsin, and elastase, could also degrade PLA.

The clear zone method using emulsified polyester agar plates was used to evaluate the population of polymer-degrading microorganisms in the environment. It was confirmed that the population of aliphatic polyester-degrading microorganisms at 30° and 50°C decreased in the order of PHB = PCL > PBS > PLA [29–31]. Suyama *et al.* [32] reported that 39 bacterial strains of class *Firmcutes* and *Proteobacteria* isolated from soil were capable of degrading aliphatic polyesters such as PHB, PCL, and PBS, but no PLA-degrading bacteria were found. These results showed that PLA-degrading microorganisms are not widely distributed in the natural environment and thus PLA is less susceptible to microbial attack in the natural environment than other microbial and synthetic aliphatic polyesters. The biodegradability of PLA depends on the environment to which it is exposed. In human or animal bodies, it is believed that PLA is initially degraded by hydrolysis and the soluble oligomers formed are metabolized by cells. Soil burial tests show that the degradation of PLA in soil is slow and that it takes a long time for degradation to start. For instance, no degradation was observed on PLA sheets after six weeks in soil [33]. Urayama *et al.* [34] reported that the molecular weight of PLA films with different optical purity of the lactate units (100% L and 70% L) decreased by 20 and 75%, respectively, after 20 months in soil.

The degradation of polylactic acid-based films by microorganisms extracted from compost was studied in a liquid medium [35]. The application of the ASTM standard (ASTM D 5209-92) did not produce biodegradation of pieces of PLA film. With the ISO/CEN standard method (ISO/CEN 14852-1998), the percentage biodegradation after 45 days was found to be 30%. The different temperature profile of medium used in two standards seemed to be the major factor in explaining the observed differences.

Commercial lipases were examined for their degradation efficiency of aliphatic polyester films in special emphasis on PLA [36]. Polyester films were immersed during 100 days in lipase solutions at 37°C at pH 7.0. Poly(butylene succinate-*co*-adipate) (PBSA) and poly(ε-caprolactone) (PCL) films were rapidly degraded during 4–17 days when either Lipase Asahi derived from *Chromobacterium viscosum*, Lipase F derived from *Rhizopus niveus* was used. Lipase Asahi could also degrade PBS film within 17 days. Lipase F-AP15 derived from *Rhizopus orizae* could degrade PBSA in 22 days. Lipase PL isolated from *Alcaligenes* sp. revealed its higher degradation activity of PLA film. PLA degraded completely at 55°C, pH 8.5 with lipase PL during 20 days. Based on the results of GPC and HPLC analyses, it was concluded that complete degradation of PLA resulted from two processes. First, the chemical hydrolysis from PLA into oligomers at higher pH and/or under higher temperature conditions, because polyesters are generally not stable under such conditions. Second, the enzymatic hydrolysis from oligomers to the monomer.

6.6.2 Biodegradation of polyhydroxyalkanoates – PHA

Degradation mechanisms

The bacterially produced poly(hydroxyalkanoates) (PHAs) are fully biodegradable in both anaerobic and aerobic conditions, and also at a slower rate in marine environments.

Poly(hydroxyalkanoates) are quite resistant to moisture, but they are rapidly biodegraded by a wide range of microorganisms [37]. The rate of enzymatic degradation of PHB and PHBV by PHA depolymerases was from two to three orders of magnitude faster than the rate of simple hydrolytic degradation. The enzymatic hydrolysis of PHB and PHV copolymers is a heterogeneous erosion process proceeding from the surface, where polymer chains are degraded initially by *endo*-scissions (randomly throughout the chain) and then by *exo*-scissions (from the chain ends) [37]. This results in subsequent surface erosion and weight loss. The average molecular weight and molecular weight distribution do not change during the enzymatic degradation because of selective degradation only at the surface, together with removal and dissolution of low molecular weight degradation products from the polymer matrix into the surrounding environment. It was reported that in the initial stages of degradation only amorphous material was consumed. Later, however, both amorphous and crystalline regions were degraded without preference.

The biodegradable properties of Biopol, thermoplastic copolyester PHBV composed of HB units and between 0 and 30% HV units, incorporated randomly throughout the polymer chain, were discussed by Byrom [38]. Biopol biodegrades in microbially active environments. Biodegradation is initiated by the action of microorganisms growing on the surface of the polymer. Microorganisms that degrade Biopol include species of *Aspergillus*, *Streptomyces*, *Actinomyces*, and *Pseudomanas*. These microorganisms secrete extracellular enzymes, such as depolymerases and esterases, that solubilize the polymer in the immediate vicinity of the cell. The soluble degradation products are then absorbed through the cell wall and metabolized to CO_2 and H_2O under aerobic conditions. The rate of degradation is dependent on a number of factors. Particularly important are the level of microbial activity (determined by the moisture level, nutrient supply, temperature, and pH) and the surface area of the polymer. A series of tests was carried out in which Biopol was composted together with "biorefuse". A weight loss of 80% was observed after 15 weeks, under these conditions when the stack was turned.

Degradation in compost

Poly-β-hydroxybutyrate/valerate copolymer (Biopol) was used as test material and cellulose powder as a reference material in a ring laboratory controlled composting test [39]. A laboratory method was presented for investigating the biodegradation of an organic test material in an aerobic composting system based on the evolution of carbon dioxide. The test becomes a basis of a European standard in connection with determining the compostability of packaging and packaging materials. The mean degree of Biopol biodegradation was 88% in comparison with 84% for microcrystalline cellulose powder.

The compost activity of poly(β-hydroxybutyrate) and a copolymer of 20% β-hydroxyvalerate was studied in a simulated municipal solid waste compost test at a constant temperature of 55°C and a constant moisture content of 54% [40]. Biodegradation was measured through weight loss and normalized for thickness. The compost activity was found to be divided into three stages with the maximum rate of polymer degradation occurring between the tenth and fifteenth day. The biodegradation rate of the valerate copolymer was seen to be much higher than that of the homopolymer.

The effect of abiotic factors such as water and air on the degradation of poly(3-hydroxybu-tyrate-*co*-3-hydroxyvalerate) (PHBV) in a compost was investigated using simulated and natu-ral environments [41]. The results showed that during a period of 50 days, water and air have little or no effect on the degradation of PHBV in garden waste compost. It was suggested that the degradation was due to microbial action only.

Changes in physical and mechanical properties of poly(hydroxybutyrate-*co*-hydroxyvalerate) during degradation in a composting medium were studied by Luo *et al.* [42]. FTIR-ATR spec-tra of the control and partly degraded PHBV specimens as a function of composting time are presented in Fig. 6.8. No detectable changes between the spectra of control and composted specimens were observed. Figure 6.9 presents typical stress vs strain plots of control and composted PHBV specimens. The ultimate tensile strength and the strain at ultimate tensile strength decreased significantly as a function of composting time. The results from the analy-sis of weight loss, SEM, molecular weight, FTIR, DSC and tensile testing suggested that the degradation of PHBV in compost medium was enzymatic rather than hydrolytic and occurred from surface and the degraded material leached out.

The biodegradation of poly-β-(hydroxybutyrate) (PHB) and poly-β-(hydroxybutyrate-*co*-β-valerate) (PHBV) was assessed by the loss of mass, tensile strength and roughness of the polymer [43]. Both polymers showed similar biodegradation in soil composting medium at 46°C and at room temperature (24°C) and in a soil simulator. After aging in soil composting medium at 46°C for 86 days, both polymers showed a decrease in the tensile strength at break (76% for PHB and 74% for PHBV). In agreement with this, the roughness of both polymers increased faster in soil composting medium at 46°C. Surface damage can be assessed by the

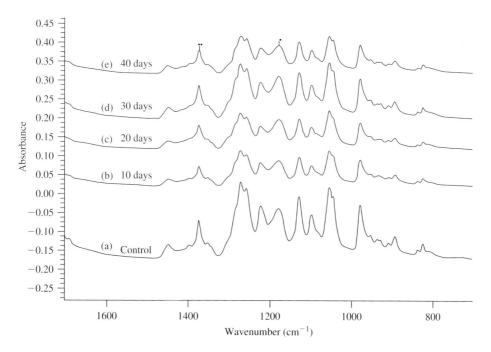

Figure 6-8 FTIR-ATR spectra of PHBV films exposed to the composting medium for different periods of time: (a) 0, (b) 10, (c) 20, (d) 30, and (e) 40 days. Reprinted with permission from [42].

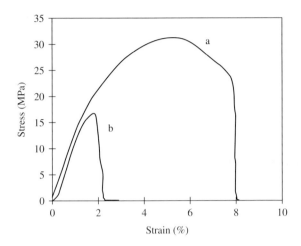

Figure 6-9 Typical stress vs. strain of control and partially degraded PHBV specimens: (a) control, (b) composted for 30 days. Reprinted with permission from [42].

measuring the surface roughness, a technique commonly used in mechanical engineering. It was suggested that roughness may be a useful parameter for evaluating the biodegradation of polymers.

The effect of temperature on the biodegradation of poly-β-(hydroxybutyrate (PHB), poly-β-(hydroxybutyrate-*co*-β-valerate) (PHBV) and poly(ε-caprolactone) (PCL) was assessed based on the mass retention when the polymers were incubated in soil compost at 46°C and 24°C [44]. Biodegradation was greatest at 46°C for the three polymers studies. PHB and PHBV showed similar biodegradation at both temperatures. PHB and PHBV were totally degraded after 104 days of aging in soil compost at 46°C and PCL degraded by 36% in 120 days. Degradation of the polymers at room temperature (24°C) was relatively slow, with losses of 51% and 56% for PHB and PHBV, respectively, after 321 days of aging. In contrast, PCL showed no biodegradation at room temperature after almost 300 days.

The effect of thermal ageing on the degradation of PHB, PHBV and PCL in soil compost-age was studied by Rosa *et al.* [45]. The biodegradability of poly-β-(hydroxybutyrate) (PHB), poly-β-(hydroxybutyrate-*co*-β-valerate) (PHBV) and poly(ε-caprolactone) (PCL) was examined following thermal aging in an oven for 192, 425 and 600 h. Different temperatures, 100, 120 and 140°C for PHB and PHBV and 30, 40 and 50°C for PCL, were used to assess the influence of this parameter on biodegradation. Thermal aging increased the biodegradability only for PHB at 120 and 140°C.

Bacterial thermoplastic polyesters poly(3-hydroxyalkanoate) (PHAs), produced by the fermentation of renewable materials, such as sugars or molasses, i.e. PHB and a copolymer of PHB(88%)/PHV(12%), were mixed with other biodegradable materials (additives) to improve their mechanical properties [46]. Plasticizers, glycerol, tributyrin, triacetin, acetyltriethylcitrate, acetyltributylcitrate, and a nucleation agent, saccharin, were used. Lubricants were glycerol-monostearate, glyceroltristearate, 12-hydroxystearate and 12-hydroxystearic acid. The biodegradability of blends was investigated in the aerobic test, under compost conditions in soil and in river water. It was found that the blends were degraded more easily in the aerobic test, i.e. in the river water and compost, than in the soil.

Several types of biodegradable medium-chain-length polyhydroxyalkanoates (mcl-PHAs) were produced by *Pseudomonas putida* KT2442 at pilot and laboratory scales from renewable long-chain fatty acids and octanoic acid [47]. All purified polymers were subjected to *in vitro* aerobic biodegradation using a compost isolate. The extent of mineralization varied from 15 to 60% of the theoretical biochemical oxygen demand (ThBOD). The polymer weight loss after 32 days ranged from 40 to 90% for the different mcl-PHAs.

Degradation in other environments
PHAs are degraded upon exposure to soil, compost, or marine sediment [48]. Biodegradation is dependent on a number of factors such as microbial activity of the environment, and the exposed surface area, moisture, temperature, pH, and molecular weight. Biodegradation of PHA under aerobic conditions results in carbon dioxide and water, whereas in anaerobic conditions the degradation products are carbon dioxide and methane. PHAs are compostable over a wide range of temperatures, even at a maximum of around of 60°C with moisture levels at 55%. Studies have shown that 85% of PHAs were degraded in seven weeks. PHAs have been reported to degrade in aquatic environments (Lake Lugano, Switzerland) within 254 days even at temperatures not exceeding 6°C.

Effective PHA destructors include various bacteria from widespread soil and water genera (*Pseudomonas, Alcaligenes, Comamonas, Streptomyces, Ilyobacter*), as well as fungi (*Ascomycetes, Basidiomycetes, Deuteromycetes, Mastigiomycetes, Myxomycetes*) [49].

The degradation dynamics of polyhydroxyalkanoates of different compositions (a PHB homopolymer and a PHB/PHV copolymer with 14 mol% of hydroxyvalerate) have been studied in a eutrophic storage reservoir for two seasons. It has been shown that the biodegradation of polymers under natural conditions depends not only on their structure and physicochemical properties but also, to a great extent, on a complex of weather-climatic conditions affecting the state of the reservoir ecosystem.

6.6.3 Biodegradation of thermoplastic starch – TPS
The suitability of an *in vitro* enzymatic method for assaying the biodegradability of starch-based materials was evaluated [50]. The materials studied included commercial starch-based materials and thermoplastic starch films prepared by extrusion from glycerol and native potato starch, native barley starch, or crosslinked amylomaize starch.

In order to verify the response of the controlled composting test method (i.e. the ISO/DIS 14855:1997, the ASTM D 5338-92) to starch at different concentrations, the maximum amount prescribed by the test method (100 g) and lower amounts (60 and 30 g), as if starch were a coingredient in a blend, were tested [51]. After 44 days of incubation (at a constant temperature of 58°C) the biodegradation curves were in a plateau phase, displaying the following final (referred to a nominal starch initial amount of 100 g): starch 100 g, 97.5%; starch 60 g, 63.7%; and starch 30 g, 32.5%. The data showed a CO_2 evolution roughly equal, in each case, to the theoretical maximum, indicating a complete starch mineralization. The average biodegradation of cellulose turned out to be 96.8% after 47 days.

The degradation of starch- and polylactic acid-based plastic films by microorganisms extracted from compost was studied in a liquid medium [52]. The various degradation products produced (carbon dioxide, biomass formed by abstraction of some of the material's carbon, soluble organic compounds, and possibly non-degraded material) were measured throughout the duration of the experiment, and total carbon balances were estimated. The experiments were conducted according to ASTM and ISO/CEN standards and used two different physical states of the material, i.e. film and powder forms. The final mineralization percentage (Cg) of starch-based material was always greater than 60%, the minimum assigned value for a biodegradable material. Moreover, the percentage of biodegradation, defined as the sum of the mineralization (Cg) and bioassimilation (Cb) was between 82 and 90%. It was concluded that for an easily biodegradable material as starch, the evolution of the way carbon repartitioned between different degradation products was quite similar whatever the experimental condition or the type of substrate. On the other hand, for a resistant material (polylactic-based plastic) exposed to these microorganisms, the nature of the biodegradation depended strongly on the experimental conditions.

6.6.4 *Biodegradation of other compostable polymers from renewable resources*

Biodegradation of cellulose
Thin-layer chromatography (TLC) grade cellulose is used as positive reference material during compostabilty studies according to international standards, e.g. ISO 14855. It was reported that the average biodegradation of cellulose during controlled composting method turned out to be 96.8 ± 6.7 (SD) after 47 ± 1 days [53].

Most of the cellulolytic microorganisms belong to eubacteria and fungi, even though some anaerobic protozoa and slime moulds able to degrade cellulose have also been described [54]. Cellulolytic microorganisms can establish synergistic relationships with non-cellulolytic species in cellulosic wastes. The interactions between both populations lead to complete degradation of cellulose, releasing carbon dioxide and water under aerobic conditions, and carbon dioxide, methane and water under anaerobic conditions.

Microorganisms capable of degrading cellulose produce a battery of enzymes with different specificities, working together. Cellulases hydrolyse the β-1,4-glycosidic linkages of cellulose. Traditionally, they are divided into two classes referred to as endoglucanases and cellobiohydrolases. Endoglucanases (endo-1,4-β-glucanases) (EGs) can hydrolyse internal bonds (preferably in cellulose amorphous regions) releasing new terminal ends. Cellobiohydrolases (exo-1,4-β-glucanases) (CBHs) act on the existing or endoglucanase-generated chain ends. Both enzymes can degrade amorphous cellulose but, with some exceptions, CBHs are the only enzymes that efficiently degrade crystalline cellulose. CBHs and EGs release cellobiose molecules. An effective hydrolysis of cellulose also requires β-glucosidases, which break cellobiose releasing two glucose molecules.

In 1999 there was considerable confusion regarding the biodegradation potential of cellulose esters [55]. There was a great deal of literature indicating that cellulose acetate (CA) above a degree of substitution (DS) of approximately 1.0 was not biodegradable while other reports suggested that CA might indeed be biodegradable. Since 1992, there have been several reports, which clearly demonstrate that CA having a DS of less than approximately 2.5 is inherently biodegradable. The general finding has been that as the DS of the CA decreases, the rate of biodegradation increases. Below a DS of ca. 2.1, degradation rates of CA in composting environments approached or exceeded those of many other known biodegradable polymers. Regarding cellulose esters with longer side chains, it has been shown that cellulose propionates (CP) below a DS of ca. 1.85 are also potentially useful as biodegradable polymers. In general, as the DS and the length of the acyl side group decreases, the rate of biodegradation increases.

A series of cellulose acetate films, differing in degree of substitution, was evaluated in the bench-scale composting system [56]. Commercially available biodegradable polymers such as poly(hydroxybutyrate-*co*-valerate) (PHBV) and polycaprolactone (PCL) were included as points of reference. Based on film disintegration and on film weight loss, cellulose acetates having a degree of substitution less than approximately 2.20 composted at rates comparable to that of PHB. NMR and GPC analyses of composted films indicated that low molecular weight fractions were removed preferentially from the more highly substituted and slower degrading cellulose acetates.

The biodegradability of cellulose acetate (CA) films with degree of substitution (DS) values of 1.7 and 2.5 using laboratory-scale compost reactors maintained at a 60% moisture content and 53°C [57]. It was found that the CA films (thickness values of 0.013 to 0.025 and 0.051 mm, respectively) had completely disappeared by the end of 7- and 18-day exposure periods, respectively. Moisture conditions in the laboratory-scale compost reactors were found to have a profound effect on the extent of CA film weight loss as a function of the exposure time. It was determined that for moisture contents of 60, 50, and 40% the time for complete CA DS-1.7 film disappearance was 6, 16, and 30 days, respectively.

The biodegradability of cellulose ester derivatives using a degradation assay based on commercially available cellulolytic enzyme preparations was found to depend on two factors: degree of substitution (DS) and substituent size [58]. The cellulose esters had acyl substituents ranging in size between propionyl and myristyl and DS values between 0.1 and nearly 3. The smaller the substituent, the higher the DS that can be tolerated by cellulolytic enzymes.

Blends of cellulose acetate having a degree of substitution of 2.49 with a cellulose acetate having a DS of 2.06 were examined [59]. Bench-scale simulated municipal composting confirmed the biodestructurability of these blends and indicated that incorporation of a plasticizer (poly(ethylene glycol)) accelerated the composting rates of the blends. *In vitro* aerobic biodegradation testing involving radiochemical labelling conclusively demonstrated that both the lower DS cellulose acetate and the plasticizer significantly enhanced the biodegradation of the more highly substituted cellulose acetate.

Several samples of cellulose acetate polymers with varying degrees of substitution (DS) between 0.7 and 1.7 have been prepared and tested for their biodegradation potential [60]. The degree of substitution (DS) of CA, i.e. the average number of acetyl groups per anhydroglucose unit, can range from 0 in the case of cellulose to 3 for the triacetate. It was found that the DS was a very significant factor in the biodegradation of these polymers. The lower the DS the easier the biodegradation. The higher DS polymers were amorphous, and the crystallinity increased with decreasing DS.

The biodegradation behaviour of the chemically modified cellulose fibres from flax was investigated by using previously isolated cellulolytic bacterial strains [61]. The extent of biodegradation of acetylated fibres, evaluated from the weight per cent remaining after 13 days of exposure to previously isolated cellulolytic bacteria *Cellvibrio* sp., decreased with increasing acetylation degree. After biodegradation the fibres showed a higher acetyl content than before the experiment, indicating that the bacteria preferentially biodegraded unsubstituted cellulose, though also acetylated chains were cleaved.

Biodegradation of chitosan

Blends of poly(3-hydroxybutyric acid) (PHB) with chitin and chitosan biodegraded in an environmental medium [62]. PHB and all blends showed high biodegradability, over 60%. The PHB/α-chitin blend containing 25% PHB degraded much faster than the pure PHB or pure α-chitin. This acceleration of the biodegradation is supposed to have arisen from the lowered crystallinity of PHB. The pure chitosan film showed slower biodegradation compared to the other films. The biodegradability of the PHB/chitosan systems was found to be significantly improved.

Biodegradation of proteins

Composting technique has been utilized to characterize the biodegradation of soy protein isolate (SPI)-based resin sheets with different additives [63]. Two different additives, i.e. Phytagel (the product of bacterial fermentation, composed of glucuronic acid, rhamnose and glucose) and stearic acid were incorporated in order to improve mechanical properties of the SPI resin. The SPI resin containing stearic acid degraded at a slower rate than the SPI resin, whereas SPI containing Phyotogel degraded at the slowest rate. Based on the spectroscopic analysis and differential scanning calorimetry studies, it was found that stearic acid and Phytagel were among the main residues in the modified SPI resins after composting. It was shown that the SPI resin degraded readily with 93.8% weight loss during the first 21 days of composting.

The effects of technological treatments of wheat gluten bioplastics on their biodegradation and on the formation of possible toxic products were studied [64]. To this end cast, hot-moulded, and mixed gluten materials were investigated with a biodegradation test in liquid culture (modified Sturm test) and in farmland soil. All gluten materials were fully degraded after 36 days in aerobic fermentation and within 50 days in farmland soil. The tests of microbial inhibition experiments revealed no toxic effects of modified gluten or of its metabolites. Thus, it was concluded that the protein bulk of wheat gluten materials was non toxic and fully biodegradable, whatever the technological process applied.

The chemiluminescence technique was used to study gelatine samples hydrolytically degraded under sterilization conditions and exposed to bacterial and fungal degradations [65]. It was found that the hydrolytic degradation mechanism was through a cleavage of the peptide bond of the protein without significant oxidation of the material. In contrast, biodegradation by bacteria and fungi at low temperatures decreased the molecular weight of the gelatine (viscosity) by the enzymatic activity but, also, produced an important oxidation in the material due to the reactive oxygen species generated in the microbial metabolism. This oxidation was detected by the drastic increase in the chemiluminescence emission of the materials. In general, much higher chemiluminescence emission intensities were observed for samples biodegraded by fungi with respect to those obtained for gelatine biodegraded by bacteria.

Proteic waste materials from pharmaceutical manufacturing, tanning and agro industries have attracted increasing attention because their intrinsic agronomic values bound to the fairly high nitrogen (12–15%) [66]. The propensity to biodegradation behaviour of casting films based on waste gelatin was investigated under incubation conditions aimed at simulating soil burial conditions. The results indicated the complete and very fast biodegradation of waste gelatin (WG) cast films. Pure WG films underwent about 60% biodegradation within 30 days of incubation. However, the negative effect of a crosslinker agent such as glutaraldehyde on the biodegradation extent and rate was observed for the films containing 1–5% crosslinking agent.

6.7. BIODEGRADATION OF BIODEGRADABLE POLYMERS FROM PETROCHEMICAL SOURCES

6.7.1 *Biodegradation of aliphatic polyesters and copolyesters*

Aliphatic polyesters and copolyesters based on succinic acid and commercialized under the name Bionolle are biodegradable in compost, in moist soil, in fresh water with activated sludge and in sea water [67].

A series of aliphatic homopolyesters and copolyesters was prepared from 1,4-butanediol and dimethyl esters of succinic and adipic acids through a two-step process of transesterification and polycondensation [68, 69]. The biodegradation of the polymers was investigated by soil burial and enzymatic hydrolysis. It was suggested that the key factor affecting material degradation was its crystallinity.

The modified Sturm test showed that poly(ethylene adipate) (PEA) and poly(butylene succinate) (PBS) were assimilated to CO_2 at a similar rate [70]. As the degree of chain branching increased, the biodegradation rate of PEA increased to a greater extent than that of PBS due to the faster reduction in the crystallinity of PEA compared to the crystallinity of PBS. Poly(alkylene succinate)s were synthesized from succinic acid and aliphatic diols with 2 to 4 methylene groups by melt polycondensation [71]. A comparative biodegradability study of the three poly(alkyl succinate)s prepared, namely poly(ethylene succinate) (PESu), poly(propylene succinate) (PPSu) and poly(butylene succinate) (PBSu), was carried out using *Rhizopus delemar* lipase. Samples having the same average molecular weight were used. The biodegradation rates of the polymers decreased following the order PPSu > PESu ⩾ PBSu and it was attributed to the lower crystallinity of PPSu compared to other polyesters, rather than to differences in chemical structure.

The bio-catalysed cleavage of ester bonds in low molecular mass model esters and aliphatic polyesters was studied [72]. The cleavage of ester bonds in liquid and solid low molecular mass model compounds by lipases exhibits substrate specificity, i.e. the cleavage rates are dependent on the chemical structure and the molecular environment the ester bonds are embedded in. In contrast, when studying the degradation of polyesters by enzymatic hydrolysis, the substrate specificity plays only a minor role. The most important quantity controlling the hydrolysis rate is the extent of mobility of the polyester chains in the crystallinity domains of the polymer. While the amorphous regions at the surface are easily degraded, the crystalline domains form a layer which protects the bulk material against enzymatic attack. Therefore, the low hydrolysis rate of the ester bonds in the crystallites is the limiting step of the overall degradation process. For aliphatic polyesters the temperature difference between the melting point of the polymer and the temperature where degradation takes place turned out to be the primary controlling parameter for polyester degradation with the lipase. If this temperature difference is less than about 30ºC, the degradation rate increases significantly.

The biodegradation and hydrolytic degradation of the high molecular weight poly(butylene succinate) homopolyester, poly(butylene adipate) homopolymer, and poly(butylene succinate-*co*-butylene adipate) copolyesters were investigated in the composting soil and NH_4Cl aqueous solutions at a pH level of 10.6 [73]. The biodegradability by microorganisms increased as the contents of butylene adipate increased, along with crystallinity and melting temperature, whereas the spherulite radius decreased. The biodegradability of poly(butylene succinate-*co*-butylene sebacate) P(BSu-*co*-BSe) and poly(butylene succinate-*co*-butylene adipate) P(BSU-co-BAd) samples, with different composition, was investigated under controlled soil burial conditions [74]. The influence of crystallinity, molar mass, chemical structure and melting temperature upon biodegradation was studied. The weight loss of poly(3-hydroxybutyrate) (PHB), of poly(3-hydroxybutyrate-*co*-3-hydroxyvalerate) 76/24 (PHBV 76/24) and of two commercial Bionolle samples, was also investigated under soil burial conditions. PHB and PHBV 76/24 showed a higher biodegradation rate than Bionolle samples but lower than some P(BSu-*co*-BSe)s and P(BSU-*co*-BAd)s. Among the homopolyesters, P(BAd) appeared more susceptible to biodegradation. P(BAd) and P(BSe) had similar melting temperature and comparable crystallinity, but the former biodegraded twice as fast as the latter. It was suggested that adipate bonds were hydrolysed faster than sebacate bonds.

The biodegradation behaviour and mechanism of aliphatic copolyester poly(butylene succinate-*co*-butylene adipate) (PBSA) by *Aspergillus versicolor* isolated from compost was studied by Zhao *et al.* [75]. Analysis of weight loss showed that more than 90% of PBSA film was assimilated within 25 days. The analyses of ^{1}H-NMR and differential scanning calorimetry (DSC) indicated that the preferred degradation took place in the adipate units and the succinate units are relatively recalcitrant to *A. verisciolor.*

The biodegradation of homopolymer poly(butylene succinate) (PBS) was studied under controlled composting conditions [76]. Composting was performed according to ISO 14855 standard at 58°C. After incubation for 90 days, the biodegradation percentage was 71.9%, 60.7%, and 14.1% for powder, film, and granule form sample, respectively. The ultimate biodegradation percentage revealed that the powder-formed sample showing the best biodegradability may be ascribed to the largest specific surface. The biodegradation process of PBS under controlled composting conditions exhibited three phases. The biodegradation in the first phase was slow (0–5 days), got accelerated in the second phase (6–66 days), and showed a levelling-off in the third phase (67–90 days). Four strains were isolated from compost and identified as *Aspergillus versicolor, Penicillum, Bacillus*, and *Thermopolyspora*. Among them, *Aspergillus versicolor* was the best PBS-degrading microorganism.

Etylene glycol/adipic acid and 1,4-butanediol/succinic acid were copolymerized in the presence of 1,2-butanediol and 1,2-decanediol to produce ethyl and n-octyl branched poly(ethylene adipate) (PEA) and poly(butylene succinate) (PBS), respectively [77]. The modified Sturm test showed that the two polymers were assimilated to CO_2 at a similar rate. As the degree of chain branching increased, the biodegradation rate of PEA increased to a greater extent than that of PBS due to the faster reduction in the crystallinity of PEA compared to the crystallinity of PBS.

Unsaturated groups were introduced into the main chains of poly(butylene succinate) (PBS) by the condensation polymerization of 1,4-butanediol with succinic acid and maleic acid (MA) [78]. The resulting aliphatic polyesters were subjected to chain extension via the unsaturated groups with benzoyl peroxide (BPO), BPO/ethylene glycol dimethacrylate, or BPO/triallyl cyanurate. Chain extension increased the glass transition temperature, decreased the melting temperature and crystallinity, and improved mechanical properties such as elongation and tensile strength. The results of the modified Sturm tests showed that the biodegradability of the unsaturated aliphatic polyesters decreased greatly because of the chain extension.

PCL- and PHB-degrading microorganisms are distributed widely and they represent 0.2 to 11.4% and 0.8% to 11.0% of the total number of microorganisms in the environment, respectively [79]. The distribution of poly(tetramethylene succinate) (PTMS)-degrading microorganisms in soil environments was quite restricted compared with the distribution of microorganisms that degrade poly(ε-caprolactone) (PCL). However, the ratios of the degrading microorganisms to the total microorganisms were almost the same for both PTMS and PCL. In soil samples in which the formation of a clear zone was observed, PTMS-degrading microorganisms constituted 0.2 to 6.0% of the total number of organisms, which was very close to the percentage (0.8 to 8.0%) observed for PCL-degrading microorganisms. Strain HT-6, an actinomycete, has good potential for treatment of PTMS, since it can degrade and assimilate various forms of PTMS, including films. It assimilated about 60% of the ground PTMS powder after eight days of cultivation.

Poly(butylene succinate-*co*-butylene adipate) (PBSA)-degrading bacterium was isolated from soil and identified as *Bacillus pumilus* [80]. It also degraded poly(butylene succinate) (PBS) and poly(ε-caprolactone) (PCL). On the other hand, poly(butylene adipate terephthalate) and poly(lactic acid) were minimally degraded by strain. The NMR spectra of degradation products from PBSA indicated that the adipate units were more rapidly degraded than 1,4-butanediol and succinate units. It was proposed to be one of the reasons why *Bacillus pumilus* degraded PBSA than PBS.

Polyesters, poly(butylene succinate adipate) (PBSA), poly(butylene succinate) (PBS), poly(ethylene succinate) (PES), poly(butylene succinate)/poly(caprolactone) blend and poly(butylene adipate terephthalate) (PBAT) were evaluated about their enzymatic degradation by lipases and chemical degradation in sodium hydroxide solution [81]. In enzymatic degradation, PBSA was the most degradable by lipase PS from *Pseudomonas* sp.; on the other hand PBAT containing aromatic ring was little degraded by 11 kinds of lipases.

The extracellular depolymerase produced by the fungus *Aspergillus fumigatus* was found to have a broad hydrolytic activity towards bacterial and synthethic aliphatic polyesters [82]. The enzyme catalysed the hydrolysis of the bacterial polyesters: poly(3-hydroxybutyrate-*co*-3-hydroxyvalerate) (PHB/HV) and poly(3-hydroxybutyrate-*co*-4-hydroxybutyrate) (P3HB/4HB), as well as synthetic polyesters: poly(ethylene adipate) (PEA), poly(ethylene succinate) (PES), poly(1,4-tetramethylene adipate) (PTMA), and commercial polyesters "Bionolle". By comparing the results of enzyme specificity experiments, degradation product analysis, and molecular modelling, it was suggested that polymer chain structure and conformation may strongly influence the activity of hydrolase toward specific polymers. Various thermophilic actinomycetes were screened for their ability to degrade a high melting point, aliphatic polyester, poly(tetramethylene succinate), at 50ºC [83]. By using the clear zone method, *Microbispora rosea*, *Excellospora japonica* and *E. viridilutea* were found to have PTMS-degrading activity. In a liquid culture with 100 mg PTMS film, *M. rosea* subsp. *Aerate* IFO 14046 degraded about 50 mg film sample after eight days.

A series of low molecular weight aliphatic biodegradable polyesters was synthesized from 1,3-propanediol and adipic acid and succinic acid and 1,4-cyclohexanedimethanediol by thermal polycondensation [84]. The biodegradability of the synthesized polyester films was tested by enzymatic degradation in phosphate buffer (pH = 7.2) in the presence of *Rhizopus delemar* lipase incubated at 37ºC, and soil burial degradation at 30ºC. The biodegradability of the polyesters depended on the crystallinity of polymers. Synthesis of high molecular weight aliphatic polyesters by polycondensation of diester with diols with and without chain extension, and the enzymatic degradation of those polyesters was investigated by Shirama *et al.* [85]. Enzymatic degradation of the polyesters was performed using three different enzymes (cholesterol esterase, lipase B, and *Rhizopus delemar* lipase) before chain extension. The enzymatic degradability varied depending on both thermal properties (melting temperature and heat of fusion (crystallinity)) and the substrate specificity of enzymes. The enzymatic degradation of chain extended polyesters was slightly smaller than that before chain extension, but proceeded steadily.

Eight polyester films derived from C_8 to C_{10} α, ω-aliphatic diols and C_4 to C_{10} dicarboxylic acids were examined to determine differences in biodegradability [86]. Two test procedures were used to evaluate degradation: agar plate cultures with a mixture of *Aspergilli*, and soil burial. In soil burial tests, weight loss of polymer from 3 to 40% was obtained after burial for one month. The order of polyester degradability in the agar culture test differed from that found in the soil burial test.

The effect of copolymer composition on the physical and thermal properties, as well as enzymatic degradation of a series of high molecular weight polyesters (butylene succinate-*co*-butylene adipate)s, was investigated [87]. The enzymatic degradation was performed in a buffer solution with *Candida cylindracea* lipase at 30ºC. The highest enzymatic degradation rate was observed for the copolyester containing 50 mol% butylene succinate units.

The filamentous fungus *Aspergillus oryzae* has been extensively used for traditional Japanese fermentation products, such as *sake* (rice wine), *shoyou* (soy sauce), and *miso* (soybean paste), for more than 1000 years [88]. This fungus could grow under culture conditions that contained emulsified poly(butylene succinate) (PBS) and poly(butylene succinate-*co*-adipate) (PBSA) as the sole carbon source, through the production of PBS-degrading enzyme in the medium, and could digest PBS and PBSA, as indicated by clearing of the culture supernatant.

6.7.2 Biodegradation of aromatic polyesters and copolyesters

Within compostable polymer materials, polyesters play a predominant role, due to their potentially hydrolysable ester bonds [89]. While aromatic polyesters such as poly(ethylene terephthalate) exhibit excellent material properties but proved to be almost resistant to microbial attack, many aliphatic polyesters turned out to be biodegradable but lack in properties, which are important for application. To combine good material properties with biodegradability, aliphatic–aromatic copolyesters have been developed. The review concerning the degradation behaviour and environmental safety of biodegradable polyesters containing aromatic constituents was given by Müller *et al.* [89].

Early investigations on the biologically induced degradation of aliphatic–aromatic copolyesters came to the conclusion that only at relatively low fractions of aromatic component can a significant degradation be observed. Later works reported that copolyesters of PET, poly(propylene terephthalate) (PPT) and PBT with adipic acid and sebacic acid, including statistical copolyesters, were degraded in a compost simulation test at 60°C up to a content of terephthalic acid of about 50 mol% [90]. Based on material properties concerns and price levels of raw materials copolyesters of 1,4-butanediol, terephthalic acid and adipic acid (BTA-copolyesters) are preferentially used for commercial biodegradable copolyesters [89]. The rate of biodegradation decreases significantly with an increasing fraction of terephthalic acid; the maximum content of terephthalic acid for BTA-materials intended to be around a maximum of 60 mol% (with regard to the acid component) [89].

The dependence of the degradation rate of BTA-copolyesters on the terephthalic acid content was investigated during degradation test on agar plates, where BTA-films were inoculated with a pre-screening mixed microbial culture from compost at 60°C [91]. Within a range of approximately 30–55 mol% terephthalic acid in the acid components such copolymers are an acceptable compromise between use properties and degradation rate.

Model oligo esters of terephthalic acid with 1,2-ethanediol, 1,3-propanediol, and 1,4-butanediol were investigated with regard to their biodegradability in different biological environments (inoculated liquid medium, soil, and compost at 60°C) [90]. SEC investigations showed a fast biological degradation of the oligomer fraction consisting of one or two repeating units, independent of the diol component used for polycondensation, while polyester oligomers with degrees of polymerization higher than two were stable against microbial attack at room temperature in a time frame of two months. At 60°C in a compost environment chemical hydrolysis also degraded chains longer than two repeating units.

Individual strains that are able to degrade aliphatic–aromatic copolyesters synthesized from 1,4-butanediol, adipic acid, and terephthalic acid were isolated by using compost as a microbial source [92]. Among these microorganisms, thermophilic actinomycetes dominate the initial degradation step. Two actinomycete strains identified as *Thermonospora fusca* exhibited high copolyester degradation rates.

Poly(butylene adipate-*co*-succinate)/poly(butylene terephthalate) copolyesters prepared by the transesterification reaction of PBAS and PBT were characterized [93]. The biodegradability of copolyesters depended on the terephthalate unit in the composition and average block length of the aromatic unit.

The dependence of the enzymatic degradation of aliphatic–aromatic copolyesters on the polymer structure was investigated by Marten *et al.* [94]. A number of defined model copolyesters containing terephthalate units as aromatic component were synthesized. It was suggested that the mobility of the polymer chains (the ability of chain segments to temporarily escape for a certain distance from the embedding crystal) is the major and general controlling factor for the biodegradability of polyesters. The results showed that the lengths of aliphatic sequences in a copolymer were not correlated with the biodegradation rate. The major factor in controlling the biodegradation rate was how highly tightly the polymer chains were fixed in the crystalline region of the material. The biodegradation rate of the copolyesters was mainly controlled by the chain mobility of the polymers, being correlated with the difference between the melting point of the polyester and the degradation temperature. The presence of longer aliphatic domains, e.g. in block copolyesters, does not facilitate the hydrolytic attack by the lipase, but longer aromatic sequences, which control the melting point of the crystalline regions, reduce the biodegradation rate. According to the authors the concept of chain mobility seems to be a quite universal way to describe and predict the biodegradation rate of synthetic polyesters, independent on their composition or microstructure.

Generally it seemed that many polyesters composed of aliphatic monomers were degradable by lipases, while most aromatic polyesters were characterized as biologically inert [95]. In aliphatic–aromatic copolyesters the tendency was found that biodegradability decreases with the content of aromatic constituents. For copolyesters composed from adipic acid, terephthalic acid and 1,4-butanediol a maximum content of about 50–60% terephthalic acid in the diacid component was reported to be the limit for biodegradability.

The model of chain mobility can generally describe the degradation behaviour of a series of polyesters with lipases such as lipase from *Pseudomonas* sp. including the missing degradability of polyesters like PET or PBT which exhibit very high melting points above 200°C [95, 96]. Recently, it was demonstrated that PET can be depolymerized by hydrolases from a new thermophilic hydrolase (TfH) *Thermobifida fusca* (former name *Thermonospora fusca*) [95, 96]. Erosion rates of 8 to 17 μm per week were obtained upon incubation at 55°C. This enzyme is especially active in degrading polyesters containing aromatic constituents and combines characteristics of lipases and esterases (activity optimum at 65°C). It was suggested that the specific modification of the active site of enzymes like TfH may open the door for enzymatic PET recycling in the future [96].

Poly(ethylene terephthalate)/copoly(succinic anhydride/ethylene oxide) copolymers (PET/PES copolymers) were synthesized by the transreaction between PET and PES [97]. The enzymatic hydrolysability by a lipase from *Rhizopus arrhizus* and biodegradability by activated sludge of the copolymers decreased with an increase in PET content. When the length of succinic acid unit in the copolymer was below 2, the hydrolysability of the copolymers decreased considerably.

6.7.3 Biodegradation of poly(caprolactone) – PCL

Polycaprolactone (PCL) is fully biodegradable when composted. The low melting point (58–60°C) of PCL makes the material suited for composting as a means of disposal, due to the temperatures obtained during composting routinely exceeding 60°C [20].

PCL degradation proceeds through hydrolysis of backbone ester bonds as well as by enzymatic attack [98]. Hence, PCL degrades under a range of conditions, biotically in soil, lake waters, sewage sludge, *in vivo*, and in compost, and abiotically in phosphate buffer solution. Hydrolysis of PCL yields 6-hydroxycaproic acid, an intemediate of the ω-oxidation, which enters the citric acid cycle and is completely metabolized.

Generally, it has been shown that the biodegradation of PCL proceeds with rapid weight loss through surface erosion with minor reduction of the molecular weight [37]. In contrast, the abiotic hydrolysis of PCL proceeds with a reduction in molecular weight combined with minor weight loss.

PCL has been shown to biodegrade in many different environments, e.g. in pure fungal cultures, in compost, in active sludge, by enzymes, and in soil [37]. It was reported that degradation of PCL in a natural environment of compost and sea water is a result of enzymatic hydrolysis and of chemical hydrolysis of the ester bonds of PCL, the dominant role in this process being played by enzymatic hydrolysis [99].

During the biodegradation of film-blown PCL, both in compost and in thermophilic anaerobic sludge, regularly spaced grooves developed on the film surface [100]. Such grooves were not seen in the cases of samples degraded in an abiotic environment. The width of the grooves increased with increasing time of biodegradation. It was interpreted as indicating preferred degradation of the amorphous part of the material. The degree of crystallinity increased from 54 to 65% during composting. Figure 6.10 shows that a shoulder was detected on the low temperature side of the main melting point in the first heating after ten days in compost. The appearance corresponds to the time of formation of the low molar mass fractions seen in the SEC chromatograms. The shoulder extended to lower temperatures with increasing degradation time. It was explained by the formation of lamellae thinner than the average thickness out of the low molar mass polymer chains formed by chain scission.

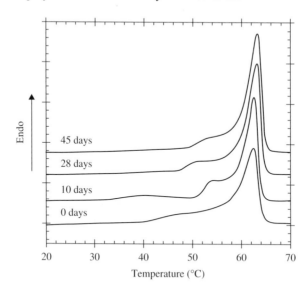

Figure 6-10 DSC curves from the first scan for the film-blown PCL degraded in compost for 0, 10, 28 and 45 days; first heating. Reprinted with permission from [100].

A series of biodegradation tests was carried out according to the standard test method 14851 in order to compare the performance of different acitvated sludge inocula on different plastic materials (polycaprolactone and starch-based material (Mater-Bi NF01U)) [101]. Cellulose was used as positive control. It was shown that the activated sludges, drawn from different wastewater treatment plants and used as inocula, had different biodegradation activities. The starch-based material was degraded to similar or higher extents than PCL with municipal sludge. Industrial sludge gave good results with both materials (PCL = 100%; starch-based material = 89%), but was less active towards cellulose. Such results raise some questions about the opportunity of also using other reference materials besides cellulose for biodegradation tests. The use of mixtures of sludges from different origins seemed to be a successful strategy to increase biodiversity and therefore increase the overall activity of inoculum.

Ammonia is the greatest nuisance odour compound among the exhaust gases that evolve during the composting process, in which raw materials with high concentrations of nitrogen, such as wastewater sludge, are decomposed [102]. A reduction of NH_3 emission during composting of wastewater sludge was tried by mixing biodegradable plastic (i.e. polycaprolactone) into composting raw material. It was found that biodegradable plastic acted as "reserve acid", i.e. it was not acid itself but degraded and released acid intermediates during the composting progress. On the basis of the results obtained, it was concluded that PCL had the characteristic of being not only compostable, but also of being able to suppress NH_3 emission during composting.

The biodegradation of polycaprolactone was examined by measuring the release of CO_2 when the plastic was mixed not with maturated compost, as in the conventional method, but with dog food used as a model fresh waste under controlled laboratory conditions [103]. From the composting in which the PCL was mixed with the dog food at various ratios, it was found that the quantity of CO_2 evolution in the presence and absence of PCL was in proportion to the PCL mixing level. The percentage of PCL decomposition, which was calculated as a ratio of the quantity of PCL decomposition to the mixing level of PCL, was 84% after 11 days in the composting using dog food, but was 59% after the same period using maturated compost.

The degradability of a biodegradable plastic depends not only on the specific kind of plastic, but also on the operational composting conditions such as temperature and the type of incoculum used. The effects of temperature and type of inoculum on the biodegradability of poly(ϵ-caprolactone) were tested in a bench-scale composting reactor under controlled laboratory composting conditions [104]. The optimum composting temperature for the PCL was found to be approximately 50°C, at which ca. 62% of the PCL was decomposed over eight days. The degradability of PCL was significantly different for each of the two types of incocula used.

The lanthanide derivatives are known as very attractive catalysts in the ring-opening polymerization of cyclic esters [105]. The influence of the lanthanides on both the hydrolytic and enzymatic degradation of the PCL obtained by ring-opening polymerization of ϵ-caprolactone with different lanthanide-based catalysts such as lanthane chloride ($LaCl_3$), ytterbium chloride ($YbCl_3$) and samarium chloride ($SmCl_3$) was assessed. Samarium seemed to slightly accelerate the hydrolytic degradation of the polymer and to slow down or inhibit its enzymatic degradation, mainly when the molecular weight of the polymer was high. The behaviour of PCL containing another lanthanide, lanthane, was dependent on the nature of the metallic ion. Complete degradation, by the lipase PS from *Pseudomonas cepacia*, was achieved only with ytterbium.

The biodegradation of electrospun nanofibres of poly(ϵ-caprolactone) (PCL) was investigated using pure-cultured soil filamentous fungi, *Asperigillus oryzae, Penicillium caseicolum, P. citrinum, Mucor* sp., *Rhizopus* sp., *Curvularia* sp., and *Cladosporium* sp. [106]. Three kinds of non-woven PCL fabrics with different mean fibre diameters (330, 360, and 510 nm) were prepared by changing the viscosities of the pre-spun PCL solutions. In the biochemical oxygen demand (BOD) test, the biodegradation of the 330 nm PCL nanofibres by *Rhizopus* sp. and *Mucor* sp. exceeded 20% and 30% carbon dioxide generation, respectively. The biodegradability of the PCL non-woven fabrics decreased with the mean fibre diameter and the 330 nm PCL nanofibre exhibited the highest biodegradability.

Polycaprolactone (PCL) powders were prepared from PCL pellets using a rotation mechanical mixer [107]. PCL powders were separated by sieves with 60 and 120 meshes into four classes: 0–125 μm, 125–250 μm, 0–250 μm and 250–500 μm. Biodegradation tests of PCL powders and cellulose powders in an aqueous solution at 25°C were performed using the coulometer according to ISO 14851. Biodegradation tests of PCL powders and cellulose powders in controlled compost at 58°C were performed according to ISO 14855-1 and by using the Microbial Oxidative Degradation Analyser (MODA) instrument according to ISO/DIS 14855-2. PCL powders were biodegraded more rapidly than cellulose powders. The reproducibility of biodegradation of PCL powders was excellent. Differences in the biodegradation of PCL powders with different classes were not observed by the ISO 14851 and ISO/DIS 14851-2. An enzymatic degradation test of PCL powders with different classes was studied using an enzyme of *Amano lipase* PS. PCL with smaller particle size was degraded more rapidly by the enzyme. PCL powders with regulated sizes from 125 μm to 250 μm were proposed as a reference material for the biodegradation test.

6.7.4 Biodegradation of poly(esteramide)s

Polyesteramides can be hydrolytically degraded through ester bond cleavages [108]. The degradation process is clearly accelerated at high temperatures, or in acid or basic pH media. In the same way, the polymer is susceptible to enzymatic attack with protease such as proteinase K.

Degradation of poly(esteramide)s differing in the amide/ester ratio under different media (water at 70°C, acid or enzymatic catalysis at 37°C) have been studied by evaluating the changes in intrinsic viscosity, in the NMR spectra and in the surface texture of samples [109]. Results indicated that the amide/ester ratio had to be lower than certain values in order to obtain samples with a high susceptibility to enzymatic catalysis. Enzymes with a protease activity appeared more effective than those with only an esterase activity.

The influence of substitution of adipic acid by terephthalic acid units on degradability under different media of poly(esteramide)s were investigated by Lozano *et al.* [110]. The degradation rate decreased with the aromatic content in aqueous media as well as in those with acid or enzymatic (protease K) catalysis.

Two types of aliphatic poly(esteramide)s were subjected to microbial degradation in basal mineral salt broth, under the attack of a yeast, *Cryptococcus laurentii*, at 20°C [111]. The first type of PEA was made by anionic ring-opening copolymerization of ε-caprolactone and ε-caprolactam, whereas the second one was synthesized by a two-step polycondensation reaction of hexanediol-1,6, hexanediamine-1,6 and adipoyl chloride. These copolymers were found to be readily degradable under biotic conditions, based on weight loss, GPC, NMR spectroscopy, and tensile property measurements. Furthermore, NMR spectroscopic analysis proved that the biodegradation of poly(esteramide)s involved the enzymatic hydrolysis of ester groups on the backbones of polymers into acid and hydroxyl groups. No breakdown of amide bonds was observed under the given biotic conditions.

Degradability of aliphatic poly(esteramide) derived from L-alanine has been studied in different media [112]. The poly(esteramide) showed a hydrolytic degradation that took place through the ester linkage and an enzymatic degradation that strongly depended on the type of enzyme. Thus, proteolytic enzymes such as papain and proteinase K were the most effective ones. Biodegradation by microorganisms from soil and activated sludges has also been evaluated.

BAK 1095, commercial polyesteramide based on caprolactam, butanediol and adipic acid was found to be completely biodegradable according to German compostability standard DIN 54900 [113]. Biodegradability of laboratory synthesized poly(esteramide) was studied in the controlled composting test according to EN 14046 standard [114]. It was found that poly(esteramide) meets the biodegradation criteria of the standard.

In order to establish the relationship between hydrophilicity and biodegradability of the aliphatic polyesters the amide group was introduced to the biodegradable aliphatic polyester

[115]. The effect of surface hydrophilicity was induced from the amide units in the polyestera-mide. Biodegradability was evaluated from various methods including activated sludge test, enzyme hydrolysis, and soil burial test. It was found that the introduction of amide groups to the aliphatic polyester improved the biodegradability, although the increase of biodegradation rate was not directly proportional to the amide content. The biodegradability of aliphatic poly-esters increased with the addition of amide functionality.

6.7.5 Biodegradation of poly(vinyl alcohol)

Poly(vinyl alcohol) (PVA) has been considered to be a truly biodegradable synthetic poly-mer since the early 1930s [116–118]. Since 1936, it was observed that PVA was susceptible to ultimate biodegradation when submitted to the action of *Fusarium lini* [118]. Suzuki and Watanabe proposed two similar degradation pathways by using different *Pseudomonas* strains [116]. In both cases the polymer is oxidized by oxidase-type enzymatic systems with evolution of hydrogen peroxide and oxygen consumption; the result of this enzymatic attack is the pro-duction of carbonyl groups along the polymer chain. Activated β-diketones or α-keto groups are subsequently hydrolysed with fission of the polymer carbon backbone [116].

The dependence of PVA biodegradation on several structural parameters, such as molecular weight, degree of saponification, and head-to-head junctions, was assessed in the presence of a selected PVA-degrading mixed culture and of the culture supernatant derived therefrom [117]. Respirometric tests carried out in the presence of selected microbial populations evidenced a limited but significant delay in the mineralization profile depending upon the degree of PVA hydrolysis, whereas no remarkable effect by molecular weight was detected. PVA is recognized as one of the very few vinyl polymers soluble in water that is also susceptible to ultimate biodeg-radation in the presence of suitably acclimated microorganisms. Nevertheless, the occurrence of specific PVA-degrading microorganisms in the environment appears to be uncommon and in most cases strictly associated with PVA-contaminated environments [118]. Most PVA-degrading was identified as aerobic bacteria belonging to *Pseudomonas*, *Alcaligenes*, and *Bacillus* genus [118].

In solution, the major biodegradation mechanism is represented by the random endocleavage of the polymer chains [118]. The initial step is the specific oxidation of 1,3-hydroxyl groups, mediated by oxidase and dehydrogenase-type enzymes, to give β-hydroxylketone as well as 1,3-diketone moities. The latter groups are susceptible to carbon–carbon bond cleavage promoted by specific β-diketone hydrolase, giving rise to the formation of carboxyl and methyl end groups.

The ultimate biological fate of PVA appears to be largely dependent upon the kind of environ-ment it reaches [118]. Accordingly, high levels of biodegradation were observed in aqueous environ-ments. On the other hand, moderate or negligible microbial attacks were repeatedly ascertained in soil and compost environments. Different hypotheses were tentatively suggested to account for these observations, such as the absence or scarce occurrence of PVA-degrading micro-organisms in soil and compost matrices, the physical state of PVA-samples, and PVA's strong interactions with the organic and inorganic components of environmental solid matrices [118].

Biodegradation in an aqueous or soil environment very markedly depends on the microbe population present and the degradation conditions [118, 119]. It proceeds quite slowly in an unadapted environment, e.g. inoculated municipal sludge gave 13% theoretical yield of CO_2 after 21 days, merely 8–9% after 74 days in soil, 7% after 48 days in compost, with a long ini-tial lag phase of 22 days [116, 119] (Figs 6.11 and 6.12). Very moderate PVA biodegradation was also detected when using compost extract as a microbial source [118, 120].

In order to assess the effect of degree of hydrolysis (HD) on the biodegradation propen-sity of PVA, samples having a similar degree of polymerization (DPn) and noticeably different

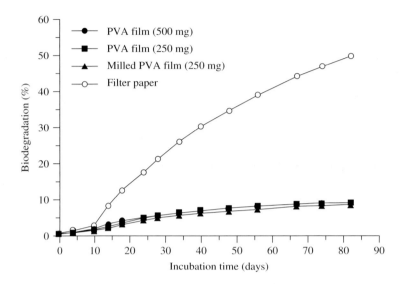

Figure 6-11 Biodegradation curves of a PVA-based film and filter paper recorded in simulated soil burial respirometric tests. Reprinted with permission from [116].

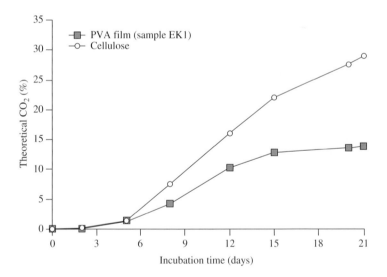

Figure 6-12 Biodegradation curves of a PVA-based film and cellulose recorded in the presence of municipal sewage sludge. Reprinted with permission from [116].

HD values were synthesized by controlled acetylation of commercial PVA (HD = 99%) and submitted to biodegradation tests in aqueous medium, mature compost and soil by using respirometric procedures [121]. Reacetylated PVA samples characterized by HD of between 25 and 75% underwent extensive mineralization when buried in solid media, while PVA (HD = 99%) showed recalcitrance to biodegradation under those conditions. An opposite trend was observed in aqueous solution, in the presence of PVA-acclimated microorganisms. In these conditions, the driving parameter affecting the microbial assimilation of PVA appeared to be water solubility of the inspected samples; the higher the solubility, the faster the biodegradation. It was suggested that biodegradation is not an absolute attribute directly related to structural features of the substrate under investigation; the conditions under which the tests are carried out have to be clearly defined.

The poly(vinyl alcohol) (PVA) degradation pathway by the enzyme from *Alcaligenes faecalis* KK314 was described by Matsumura *et al.* [122]. It was proposed that the hydroxy group of PVA was first dehydrogenated into the corresponding carbonyl group to form the β-hydroxy ketone moiety which was followed by the aldolase-type cleavage to produce the methyl ketone and the aldehyde terminals by the PVA-assimilating strain *Alcaligenes faecalis* KK314. Both the biodegradation steps of dehydrogenation and subsequent aldolase-type cleavage were catalysed by the same protein.

A mathematical model that governs the temporal change of the weight distribution with respect to the molecular weight in order to determine the enzymatic degradation rate numerically was proposed by Watanabe *et al.* [123]. As an example the GPC profiles of polyvinyl alcohol were introduced into the numerical computation. PVA was degraded by random oxidation of hydroxyl groups and following cleavage of the carbon–carbon chain between two carbonyl groups/a carbonyl group and an adjacent hydroxymethine group either by hydrolase or aldolase.

The biodegradability of PVA was investigated under different conditions by respirometric determinations, iodometric analysis, and molecular weight evaluation [124]. Microbial inocula derived from the sewage sludge of municipal and paper mill wastewater treatment plants were used. A rather active PVA-degrading bacterial mixed culture was obtained from the paper mill sewage sludge. The influence of some polymer properties such as molecular weight and degree of hydrolysis on the biodegradation rate and extent was investigated in the presence of either the acclimated mixed bacterial culture or its sterile filtrate. Kinetic data relevant to PVA mineralization and to the variation of PVA concentration, molecular weight, and molecular weight distribution revealed a moderate effect of the degree of hydrolysis.

The rates and extents of absorption and desorption of polyvinyl alcohol (PVA) samples on different solid substrates comprising montmorillonite, quartz sand, and farm soil, as well as humic acid mixture were studied [125]. Biodegradation experiments carried out in liquid cultures of PVA adsorbed on montmorillonite showed that mineralization of the adsorbed PVA was much lower than that detected for the non-adsorbed PVA. It was suggested that irreversible adsorption of PVA on the clay component occurred in soil, thus substantially inhibiting PVA biodegradation.

6.8. BIODEGRADATION OF BLENDS

6.8.1 *Blends of PLA*

Biodegradability of poly(lactic acid) (PLA) and poly(lactic acid)/corn starch composites with and without lysine diisocyanate (LDI) were evaluated by enzymatic degradation using proteinase K and burial tests [126]. The addition of corn starch resulted in a faster rate of enzymatic biodegradation and the composites with LDI were more difficult to degrade than those without it. In a burial test, pure PLA was little degraded but the composites gradually degraded. The degradation of the composite without LDI was faster than that of the composite with LDI.

Two different types of biodegradable polyester composites, PLLA fibre-reinforced PCL and PCL/PLLA blend films, were prepared with a PCL/PLLA ratio of 88/12 (w/w) and their enzymatic degradation was investigated by the use of *Rhizopus arrhizus* lipase and proteinase K as degradation enzymes for PCL and PLLA chains, respectively [127]. In the fibre-reinforced film, the presence of PLLA fibres accelerated the lipase-catalysed enzymatic degradation of PCL matrix compared with that in the pure PCL film, whereas in the blend film, the presence of PLLA chains dissolved in the continuous PCL-rich domain retarded the lipase-catalysed enzymatic degradation of PCL chains. In contrast, in the fibre-reinforced film, the proteinase K-catalysed enzymatic degradation of PLLA fibres was disturbed compared with that of the pure PLLA film, whereas in the blend film, the proteinase K-catalysed enzymatic degradation rate of particulate PLLA-rich domains was higher than that of pure PLLA film.

6.8.2 Blends of PHA

Blends of poly(3-hydroxybutyrate-*co*-3-hydroxyvalerate) with corn starch were evaluated for their biodegradability in natural compost by measuring changes in physical and chemical properties over a period of 125 days [128]. The degradation of plastic material, as evidenced by weight loss and deterioration in tensile properties, correlated with the amount of starch present in the blends (neat PHBV <30% <50%). Incorporation of poly(ethylene oxide) (PEO) into starch–PHBV blends had little or no effect on the rate of weight loss. Starch in blends degraded faster than PHBV and it accelerated PHBV degradation. After 125 days of exposure to compost, neat PHBV lost 7% weight (0.056% weight loss/day), while the PHBV component of a 50% starch blend lost 41% of its weight (0.328% weight loss/day).

The degradation of atactic poly(R,S)-3-hydroxybutyrate (a synthetic amorphous analogue of natural PHB), binary blends with natural PHB and poly(L-lactic acid) (PLLA), respectively, has been investigated in soil [129]. In such a natural environment, a-PHB blend component was found to biodegrade. The degradation of a-PHB-containing blends proceeded faster than that of respective plain n-PHB and PLLA.

6.8.3 Blends of starch

Commercially available biodegradable aliphatic polyesters, i.e. having high molecular weight poly(ε-caprolactone) (PCL) and polylactide (PLA), were melt blended with polysaccharide/ starch either as corn starch granules or as thermoplastic corn starch after plasticization with glycerol [130]. Interface compatibilization was achieved via two different strategies depending on the nature of the polyester chains. In the case of PLA/starch compositions, PLA chains were grafted with maleic anhydride through a free radical reaction conducted by reactive extrusion. As far as PCL/starch blends were concerned, the compatibilization was achieved via the interfacial localization of amphiphilic graft copolymers formed by grafting of PCL chains onto a polysaccharide backbone such as dextran. Finally, the biodegradability of so-obtained PCL/starch blends has been investigated by composting. For doing so, thin films (ca. 100 μm thick) were buried in an aerated composting bin for 120 days at 25–30°C, then followed by 20 days more at a higher temperature of 35–40°C. The film weight loss increased with the starch content. The degradation started first within the starch phase and then occurred within the polyester matrix. These compatibilized PCL/starch compositions displayed much more rapid biodegradation as measured by composting testing.

The biodegradability of native and compatibilized poly(ε-caprolactone) (PCL)–granular starch blends in composting and culture conditions was studied. The inherent biodegradability of the host polyester has been shown to increase with compatibilization within the PCL–starch compositions [131]. It was observed that the weight loss during composting increased with the decrease in interfacial tension between filler and polymer. In general, it was concluded that inherent biodegradability does not depend very significantly on the concentration of starch in the polyester matrix, but on the compatibilization efficiency.

Different proportions of starch were blended with poly(β-hydroxybutyrate)-*co*-poly(β-hydroxyvalerate) (PHB-V) or poly(ε-caprolactone) (PCL) by extrusion [132]. The biodegradability of the blends in soil compost was assessed after thermal aging for 192, 425, and 600 h at different temperatures. Two temperatures were chosen for each polymer: 100°C and 140°C for PHB-V and its blends and 30°C and 50°C for PCL and its blends. The samples of PHB-V degraded more than those of PCL, because after about 62 days of aging in soil compost, the first polymer had biodegraded almost 100%. The addition of starch to PCL slightly increased the loss of mass during biodegradation. For PHB-V the addition of 50% starch made the blend more susceptible to biodegradation, with PHB-V50 totally degraded in only 33 days. For the blends prepared, only the biodegradation of PHB-V25 was affected by thermal aging.

6.8.4 Blends of PCL

Poly(ε-caprolactone) was blended with poly(butylene succinate) (PBS) (PCL/PBS = 30/70) to improve the heat stability of PCL [133]. The processability of the blended samples was improved by γ-ray irradiation. The soil degradation test showed that the blend film buried in the soil was almost degraded (97%) after two months and completely degraded after two and a half months. On the contrary, the samples placed on the surface of the soil degraded only 3.5% after four months. From these findings it was confirmed that microorganisms contribute to degradation in soil. The blend sample used as garbage bags was well degraded (almost 50%) after a two month burial test.

The effects of replacing PCL with acrylic acid grafted PCL (PCL-*g*-AA) on the structure and properties of a PCL–chitosan composite were investigated [134]. Resistance to water was higher in the PCL-*g*-AA–chitosan blend, and consequently so was its resistance to biodegradation in soil and in an enzymatic environment. Nevertheless, weight loss of blends buried in soil or exposed to an enzymatic environment indicated that both blends were biodegradable, especially at high levels of chitosan content.

Biodegradation of blends of poly(ε-caprolactone) (PCL) with poly(vinyl butyral) (PVB) blends was studied in the soil and by bacterial strains of *Bacillus subtilis* and *Escherichia coli* isolated from the soil [135]. Weight loss was observed in all the blends. PCL-rich blends showed more degradation, which was faster in the natural environment than in the laboratory. Blends in the *Bacillus subtilis* strain showed more degradation as compared to the *E. coli* strain.

Poly(ε-caprolactone) was blended with thermoplastic starch prepared from regular corn starch [136]. PCL showed no significant reduction in mass after incubation with α-amylase, whereas blends containing corn starch were more susceptible to this enzyme. The biodegradation seen in simulated soil agreed with the findings for degradation by α-amylase.

Poly(L-lactic acid) (PLLA) and poly(ε-caprolactone) (PCL), and their films blended with or without 50 wt% poly(ethylene glycol) (PEG), were prepared by solution casting [137]. Porous films were obtained by water extraction of PEG from solution-cast phase-separated PLLA-*blend*–PCL-*blend*–PEG films. Polymer blending as well as pore formation enhanced the enzymatic degradation of biodegradable polyester blends.

Modified polycaprolactone was synthesized by melt reaction of PCL and reactive monomers such as glycidyl methacrylate (GMA) and maleic anhydride (MAH) in the presence of benzoyl peroxide in a Brabender mixer [138]. Reactive blends of the PCL-*g*-GMA and the gelatinized starch with glycerin were prepared and their mechanical properties and biodegradabilities were investigated. Reactive blends of PCL-*g*-GMA and starch showed a well-dispersed starch domain in the matrix and better mechanical strength than the unmodified PCL–starch blend. However, the reaction between PCL-*g*-GMA and starch induced a crosslinking during the reactive blending and this crosslinking in the blend lowered the biodegradation of the blend during the composting test.

Biodegradable polyester blends were prepared from poly(L-lactic acid) (PLLA) and poly (ε-caprolactone) (PCL) (50/50) by melt-blending, and the effects of processing conditions (shear rate, time, and strain) of melt-blending on proteinase K- and lipase-catalysed enzymatic degradability were investigated by gravimetry, differential scanning calorimetry (DSC), and scanning electron microscopy (SEM) [139]. The proteinase K-catalysed degradation rate of the blend films increased and levelled off with increasing the shear rate, time, or strain for melt-blending, except for the shortest shear time of 60 s. It was revealed that the biodegradability of PLLA–PCL blend materials can be manipulated by altering the processing conditions of melt-blending (shear rate, time, or strain) or the sizes and morphology of PLLA-rich and PCL-rich domains.

The biodegradability properties of poly(ε-caprolactone) and modified adipate starch blends, using EDENOL-3203 (an C_{18} alkyl epoxy stearate), were investigated in the laboratory by burial tests in agricultural soil [140]. The biodegradation process was carried out using the respirometric test according to ASTM D 5988-96, and the mineralization was followed by both variables such as carbon dioxide evolution and mass loss. It was found that the presence of modified adipate starch accelerated the biodegradation rate.

6.8.5 Blends of aliphatic–aromatic copolyesters

The blends of aliphatic–aromatic copolyesters synthesized from dimethyl succinate, dimethyl terephthalate and butanediol with starch was studied by soil burial [141]. Blends of copolyesters with starch posssessed higher degradation rate but lower tensile strength as compared with unfilled copolyesters.

Biodegradation of natural and synthetic copolyesters in two different natural environments, i.e. in compost with activated sludge at a sewage farm and in the Baltic Sea, was studied by Rutkowska *et al.* [142]. The results revealed that the natural aliphatic copolyester 3-hydroxybutyrate-*co*-3-hydroxyvalerate (PHBV) and its blends with the synthetic aliphatic–aromatic copolyester of 1,4-butanediol with adipic and terephthalic acids degraded faster in compost than in seawater. In both natural environments, blends degraded faster than aliphatic–aromatic copolyester, but at a slower rate than natural component PHBV.

Biodegradability in soil of the poly(butylene succinate adipate) (PBSA)–starch films prepared with starch contents of 5–30% by weight and processed by blown film extrusion was assessed [143]. The rate of biodegradation in soil, as measured by respirometry, increased significanly as the starch content was increased to 20% and then plateaued.

6.8.6 PVA blends

Biodegradability in a typical environment medium of blend films composed of bacterial poly(3-hydroxybutyric acid) (PHB) and chemically synthesized poly(vinyl alcohol) (PVA) was investigated by BOD test [144]. Water from the River Tama (Tokyo, Japan) was used as an environmental medium. The degradation profile of the blend films was found to depend on their blend compositions. The blend films with PHB-rich composition showed higher degradation rate and higher final degradation ratio than the pure PHB film.

Hybrid blends based on poly(vinyl alcohol) (PVA) and collagen hydrolysate (CH), an abundant, added value waste product of the leather industry, have been processed by melt blow extrusion [145, 146]. Biodegradation experiments performed under anaerobic conditions evidenced a positive effect of collagen hydrolysate on the mineralization rate of PVA–CH blends. No differences in biodegradation under aerobic conditions of PVA and PVA–CH blends at 20°C were observed when an adopted inoculum (i.e. obtained from a previous PVA biodegradation test) was used [145, 146]. On the contrary, when at lower temperature (5°C) the biodegradation level of CH-free PVA films was much lower than that detected for PVA–CH blend film.

Soil burial degradation behaviour of miscible blend systems of poly(vinyl alcohol)/partially deacetylated chitin, PVA/chitin-*graft*-poly(2-methyl-2-oxazoline), and PVA/chitin-*graft*-poly(2-ethyl-2-oxazoline) was investigated in comparison with the case of a pure PVA film [147]. The rate of weight decrease in these PVA–chitin derivative hybrids was higher than that of control PVA in the soil burial test. Fourier transform infrared spectra of the recovered samples of the blends showed an apparent increase of the absorption intensity due to β-diketone structure in PVA, which reflected the progress of biodegradation of PVA by PVA-oxidizing enzymes. The triad tacticity and number-average molecular weight of PVA in the hybrids after soil burial determined by ^{1}H-NMR and size exclusion chromatography, respectively, were almost the same as those before soil burial. It was suggested that enzymatic degradation of the hybrid films occurred mainly on the surface and that degradation of the PVA-based samples in the soil was accelerated by blending the chitin derivatives.

The effects of addition of the hydrophilic water-insoluble PVA on the non-enzymatic and enzymatic hydrolysis of hydrophobic PLLA were investigated [148]. The results of gravimetry, gel permeation chromatography (GPC), differential scanning calorimetry (DSC), tensile testing, and scanning electron microscopy (SEM) exhibited that the non-enzymatic and enzymatic hydrolysis of PLLA was accelerated by the presence of PVA and both the hydrolysis rates increased dramatically with a rise in PVA content in the blend films. The enhanced non-enzymatic hydrolysis of PLLA in the blend films was ascribed to the increased water concentration around PLLA molecules and water supply rate to them by the presence of hydrophilic PVA both in PLLA-rich and PVA-rich phases. However, the accelerated enzymatic hydrolysis of PLLA in the blend films was due to occurrence of enzymatic hydrolysis at the interfaces of PLLA-rich and PVA-rich phases inside the blend films as well as at the film surfaces.

The main shortcomings of biodegradable starch/poly(vinyl alcohol) (PVA) film are hydrophilicity and poor mechanical properties [149]. With an aim to overcome these advantages, corn starch was methylated and blend film was prepared by mixing methylated corn starch (MCS) with PVA. Enzymatic, microbiological and soil burial biodegradation results indicated that the biodegradability of the MCS/PVA film strongly depended on the starch proportion in the film matrix.

The biodegradation of PVA blends with natural polymers, such as gelatin, lignocellulosic by-products (sugar cane bagasse), as well as poly(vinyl acetate), was investigated in respirometric tests aimed at reproducing soil burial conditions [150]. The collected data evidenced that the biodegradation of PVA and PVA-based materials was rather limited under soil conditions. Additionally, PVA depresses the biodegradation of some of the investigated blends, particularly when mixed with gelatin.

The biodegradation of chitosan modified PVA–starch blends by compost was reported and compared with unmodified film by Jayasekara *et al.* [151]. Within 45 days of composting, the starch and glycerol components were fully degraded, leaving the PVA component essentially intact for unmodified blends. The film characteristics were improved by surface modification with chitosan. There was slight evidence that PVA biodegradation had been initiated in composted, surface modified starch–PVA blends.

6.8.7 Miscellaneous

Biodegradation of plastics was tested in the compost stored at $-20°C$, $4°C$ and $20°C$ for different periods [152]. It was found that biodegradation of cellulose in the compost was almost independent of the storage time and temperature. In contrast, biodegradability of both polycaprolactone (PBS) and poly(butylene succinate) (PBS) depended strongly on the storage conditions.

The degradation of poly(3-hydroxybutyrate) (PHB), a synthetic aliphatic polyester (Sky-Green) and a starch-based polymer material (Mater-Bi) was investigated in various soil types (i.e. forest soil, sandy soil, activated sludge soil and farm soil), and the characteristics of fungi that degrade those polymers were examined [153]. Biodegradation of all three polymers was most active in the activated sludge soil. In both the soil burial test and the modified Sturm test the order of the biodegradation rate was PHB > Sky-Green > Mater-Bi.

The poly(ε-caprolactone) (PCL) and poly((R)-3-hydroxybutyrate) (R-PHB) films with a hydrophilic surface were prepared by the alkali treatment of their as-cast films in NaOH solutions of different concentrations [154]. The alkali-treated PCL and R-PHB films, as well as the as-cast PCL and R-PHB films, were biodegraded in soil controlled at $25°C$. The alkali treatment enhanced the hydrophilicities and biodegradabilities of the PCL and R-PHB films in the soil. The biodegradabilities of the as-cast aliphatic polyester films in controlled soil decreased in the following order: PCL > R-PHB > PLLA, in agreement with that in controlled static seawater.

Degradabilities of four kinds of commercial biodegradable plastics, copolyester of polyhydroxybutyrate (PHB, 92%) and valerate (8%) (PHBV), polycaprolactone (PCL), blends of starch and polyvinyl alcohol (SPVA) and cellulose acetate (CA), were investigated in waste landfill model reactors that were operated anaerobically and aerobically [155]. PCL showed film breakage under both conditions, which may have contributed to a reduction in the waste volume regardless of aerobic or anaerobic conditions. Effective degradation of PHBV plastic was observed in the aerobic conditions, though insufficient degradation was observed in the anaerobic condition. In contrast, aeration may not significantly enhance the volume reduction of SPVA and CA plastics.

Aerobic and anaerobic biodegradation of four different kinds of polymers, polylactic acid, polycaprolactone, a starch–polycaprolactone blend (Mater-Bi) and poly(butylene adipate-*co*-terephthalate) (Eastar Bio), has been studied in the solid state under aerobic conditions and in the liquid phase under both aerobic and anaerobic conditions [156]. Several standard test methods (ISO 14851, ISO 14853, ASTM G 21-90 and ASTM G-22-76 and NF X 41-514) were used to determine the biodegradability. To determine the efficiency of the biodegradation of polymers, quantitative (mass variations, oxygen uptake, pressure variations, biogas generation and composition, biodegradation percentages) and qualitative (variation of T_g and T_f, variation of molar mass by SEC, characterization by FTIR and NMR spectroscopy) analyses were made and materials were characterized before and after 28 days of degradation.

Melt-pressed films of polycaprolactone (PCL) and poly(lactic acid) (PLA) with processing additives, $CaCO_3$, SiO_2, and erucamide, were subjected to pure fungal cultures *Aspergillus fumigatus and Penicillium simplicissimum* and to composting [157]. The PCL films showed a rapid weight loss with a minor reduction in the molecular weight after 45 days in *A. fumigatus*. The addition of SiO_2 to PCL increased the rate of bio(erosion) in *A. fumigatus* and in compost. PLA without additives and PLA containing SiO_2 exhibited the fastest (bio)degradation, followed by PLA with $CaCO_3$. The degradation of the PLA films was initially governed by chemical hydrolysis, followed by acceleration of the weight change and of the molecular weight reduction.

Biodegradation of poly(ε-caprolactone) (PCL), cellulose acetate (CA) and their blends using an aerobic biodegradation technique (the Sturm test) was compared [158]. The 40PCL–60CA blend showed faster biodegradation than the other blends. PCL was more susceptible to attack by a mixture of fungi on solid medium than was CA but showed a lower loss of mass than the latter polymer; the 60 PCL–40CA blend showed the greatest loss of mass during the period of evaluation. In contrast, in liquid medium, PCL showed a greater loss of mass.

6.9. SUMMARY OF COMPOSTING

Table 6.8. Biodegradation results of compostable polymer materials [159]

Polymer	Name	Company	Biodegradation mineralization, %*
Polymers based on renewable resources			
PLA	NatureWorks	Cargill Dow	100
PHBV	Biopol D400G, HV = 7%	Monsanto	100
Polymers based on petroleum resources			
PCL	CAPA 680	Solvay	100
PEA	BAK 1095	Bayer	100
PBSA	Bionolle 3000	Showa	90
PBAT	Eastar Bio 14766	Eastman	100

* At 60 days in controlled composting according to ASTM 5336.

REFERENCES

[1] Bellia G., Tosin M., Floridi G., Degli-Innocenti F.: Activated vermiculite, a solid bed for testing biodegradability under composting conditions, *Polym. Deg. Stab.* 66 (1999) 65.

[2] Degli-Innocenti F., Bellia G., Tosin M., Kapanen A., Itävaara M.: Detection of toxicity released by biodegradable plastics after composting in activated vermiculite, *Polym. Deg. Stab.* 73 (2001) 101.

[3] Longieras A., Copinet A., Bureau G., Tighzert L.: An inert solid medium for simulation of material biodegradation in compost and achievement of carbon balance, *Polym. Deg. Stab.* 83 (2004) 187.

[4] Bellia G., Tosin M., Degli-Innocenti F.: The test method of composting in vermiculite is unaffected by the priming effect, *Polym. Deg. Stab.* 69 (2000) 113.

[5] Jang J.-Ch., Shin P.-K., Yoon J.-S., Lee I.-M., Lee H.-S., Kim M.-N.: Glucose effect on the biodegradation of plastics by compost from food garbage, *Polym. Deg. Stab.* 76 (2002) 155.

[6] Yang H.-S., Yoon J.-S., Kim M.-N.: Dependence of biodegradability of plastics in compost on the shape of specimens, *Polym. Deg. Stab.* 87 (2005) 131.

[7] Jayasekara R., Lonergan G.T., Harding I., Bowater I., Halley P., Christie G.B.: An automated multi-unit composting facility for biodegradability evaluations, *J. Chem. Technol. Biotechnol.* 76 (2001) 411.

[8] Száraz L., Beczner J., Kayser G.: Investigation of the biodegradability of water-insoluble materials in a solid test based on the adaption of a biological oxygen demand measuring system, *Polym. Deg. Stab.* 81 (2003) 477.

[9] Körner I., Braukmeier J., Herrenklage J., Leikam K., Ritzkowski M., Schlegelmilch M., Stegmann R.: Investigation and optimization of composting processes – test systems and practical examples, *Waste Management* 23 (2003) 17.

[10] Ghorpade V.M., Gennadios A., Hanna M.A.: Laboratory composting of extruded poly(lactic acid) sheets, *Biores. Technol.* 76 (2001) 57.

[11] Grima S., Bellon-Maurel V., Feuilloley P., Silvestre F.: Aerobic biodegradation of polymers in solid-state conditions: a review of environmental and physicochemical parameter settings in laboratory simulations, *J. Polym. Environm.* 8 (2000) 183.

[12]   Joo S.B., Kim M.N., Im S.S., Yoon J.S.: Biodegradation of plastics in compost prepared at different composting conditions, *Macromol. Symp.* 224 (2005) 355.

[13]   Itävaara M., Vikman M.: An overview of methods for biodegradability testing of biopolymers and packaging materials, *J. Environ. Polym. Degrad.* 4 (1996) 29.

[14]   Calmon A., Silvestre F., Bellon-Maurel V., Roger J.-M, Feuilloley P.: Modelling easily biodegradability of materials in liquid medium – relationship between structure and biodegradability, *J. Environ. Polym. Deg.* 7 (1999) 135.

[15]   Pagga U., Schäfer A., Müller R.-J., Pantke M.: Determination of the aerobic biodegradability of polymeric material in aquatic batch tests, *Chemosphere* 42 (2001) 319.

[16]   Gu J.-G., Gu J.-D.: Methods currently used in testing microbiological degradation and deterioration of a wide range of polymers with various degrees of degradability: a review, *J. Polym. Environm.* 13 (2005) 65.

[17]   Kunioka M., Ninomiya F., Funabashi M.: Biodegradation of poly(lactic acid) powders proposed as the reference test materials for the international standard of biodegradation evaluation methods, *Polym. Deg. Stab.* 91 (2006) 1919.

[18]   Lunt J.: Large-scale production, properties and commercial applications of polylactic acid polymers, *Polym. Deg. Stab.* 59 (1998) 145.

[19]   Auras R., Harte B., Selke S.: An overview of polylactides as packaging materials, *Macromol. Biosci.* 4 (2004) 835.

[20]   Biodegradable Plastics – development and environmental impacts, Nolan-ITU Pty Ltd, prepared in association with ExcelPLas, Australia, October 2002, www. environment.gov.au/settlements/ publications/waste/degradables/biodegradable

[21]   Techno-economic feasibility of large-scale production of bio-based polymers in Europe (PRO-BIP), Final Report, Utrecht/Karlsruhe, October 2004.

[22]   Ho K.-L.G., Pometto III A.L., Gadea-Rivas A., Briceño, Rojas A.: Degradation of polylactic acid (PLA) plastic in Costa Rican soil and Iowa State University compost rows, *J. Environ. Polym. Deg.* 7 (1999) 173.

[23]   Kijchavengkul T., Auras R., Rubino M., Ngouajio M., Fernandez R.T.: Development of an automatic laboratory-scale respirometric system to measure polymer biodegradability, *Polymer Testing* 25 (2006) 1006.

[24]   Tuominen J., Kylmä J., Kapanen A., Venelampi O., Itävaara M., Seppälä J.: Biodegradation of lactic acid based polymers under controlled composting conditions and evaluation of the ecotoxicological impact, *Biomacromolecules* 3 (2002) 445.

[25]   Itävaara M., Karjomaa S., Selin J.-F.: Biodegradation of polylactide in aerobic and anaerobic thermophilic conditions, *Chemosphere* 46 (2002) 879.

[26]   Kale G., Aura R., Singh S.P.: Comparison of the degradability of poly(lactide) packages in composting and ambient exposure conditions, *Packag. Technol. Sci.* 20 (2007) 49.

[27]   Khabbaz F., Karlsson S., Albertsson A.-Ch.: Py-GC/MS – an effective technique for characterizing the degradation mechanism of poly(L-lactide) in different environments, *J. Appl. Polym. Sci.* 78 (2000) 2369.

[28]   Tokiwa Y., Jarerat A.: Microbial degradation of aliphatic polyesters, *Macromol. Symp.* 201 (2003) 283.

[29]   Tokiwa Y., Calabia B.P.: Biodegradability and biodegradation of poly(lactide), *Appl. Microbiol. Biotechnol.* 72 (2006) 244.

[30]   Pranamuda H., Tokiwa Y., Tanaka H.: Polylactide degradation by an *Amycolatopsis* sp., *Appl. Environ. Microbiol.* 63 (1997) 1637.

[31]   Tansengco M.L., Tokiwa Y.: Comparative population study of aliphatic polyester-degrading microorganisms at 50°C, *Chem. Lett.* 27 (1998) 1043.

[32]   Suyama T., Tokiwa Y., Ouichanpagdee P., Kanagawa T., Kamagata Y.: Phylogenetic affiliation of soil bacteria that degrade aliphatic polyesters available commercially as biodegradable plastics, *Appl. Environ. Microbiol.* 64 (1998) 5008.

[33] Ohkita T., Lee S.H.: Thermal degradation and biodegradability of poly(lactic acid)/corn starch bio-composites, *J. Appl. Polym. Sci.* 100 (2006) 3009.

[34] Urayama H., Kanamori T., Kimura Y.: Properties and biodegradability of polymer blends of poly(L-lactide) with different optical purity of the lactate units, *Macromol. Mater. Eng.* 287 (2002) 116.

[35] Gattin R., Copinet A., Bertrand C., Couturier Y.: Comparative biodegradation study of starch- and polylactic acid-based films, *J. Polym. Environm.* 9 (2001) 11.

[36] Hoshino A., Isono Y.: Degradation of aliphatic polyester films by commercially available lipases with special reference to rapid and complete degradation of poly(L-lactide) film by lipase PL derived from *Alcaligenes* sp., *Biodegradation* 13 (2002) 141.

[37] Hakkarainen M.: Aliphatic polyesters: abiotic and biotic degradation and degradation products, *Adv. Polym. Sci.* 157 (2002) 115.

[38] Byrom D.: The synthesis and biodegradation of polyhydroxyalkanoates from bacteria, *Int. Biodeter. Biodegrad.* 31 (1993) 199.

[39] Pagga U., Beimborn D.B., Boelens J., De Wilde B.: Determination of the aerobic biodegradability of polymeric material in laboratory controlled composting test, *Chemosphere* 31 (1995) 4475.

[40] Yue C.L., Gross R.A., McCarthy S.P.: Composting studies of poly(β-hydroxybutyrate-co-β-valer-ate), *Polymer Deg. Stab.* 51 (1996) 205.

[41] Eldsätter C., Karlsson S., Albertsson A.-Ch.: Effect of abiotic factors on the degradation of poly(3-hydroxybutyrate-*co*-3-hydroxyvalerate) in simulated and natural composting environments, *Polym. Deg. Stab.* 64 (1999) 177.

[42] Luo S., Netravali A.N.: A study of physical and mechanical properties of poly(hydroxybutyrate-*co*-hydroxyvalerate) during composting, *Polym. Deg. Stab.* 80 (2003) 59.

[43] Rosa D.S., Lotto N.T., Lopes D.R., Guedes C.G.F.: The use of roughness for evaluating the biodegradation of poly-β-(hydroxybutyrate) and poly-β-(hydroxybutyrate-co-β-valerate), *Polymer Testing* 23 (2004) 3.

[44] Lotto N.T., Calil M.T., Guedes C.G.F., Rosa D.S.: The effect of temperature on the biodegradation test, *Materials Science and Engineering* 24 (2004) 659.

[45] Dos Santos Rosa D., Callil M.R., Das Graças Fassina Guedes C., Rodrigues T.C.: Biodegradability of thermally aged PHB, PHBV, and PCL in soil compostage, *J. Polym. Environm.* 12 (2004) 239.

[46] El-Hadi A., Schnabel R., Straube E., Müller G., Henning S.: Correlation between degree of crystallinity, morphology, glass temperature, mechanical properties and biodegradation of poly(3-hydroxyalkanoate) PHAs and their blends, *Polymer Testing* 21 (2002) 665.

[47] Kellerrhals M.B., Kessler B., Witholt B., Tchouboukov, Brandl H.: Renewable long-chain fatty acids for production of biodegradable medium-chain-length polyhydroxyalkanoates (mcl-PHAs) at laboratory and pilot plant scales, *Macromolecules* 33 (2000) 4690.

[48] Reddy C.S.K. Ghai R., Rashmi, Kalia V.C.: Polyhydroxyalkanoates: an overview, *Bioresource Technology* 87(2003) 137.

[49] Volova T.G., Gladyshev M.I., Trusova M.Yu., Zhila N.O.: Degradation of polyhydroxyalkanoates and the composition of microbial destructors under natural conditions, *Microbiology* 75 (2006) 593.

[50] Vikman M., Itävaara M., Poutanen K.: Measurement of the biodegradation of starch-based materials by enzymatic methods and composting, *J. Envir. Polym. Deg.* 3 (1995) 23.

[51] Degli-Innocenti F., Tosin M., Bastioli C.: Evaluation of the biodegradation of starch and cellulose under controlled composting conditions, *J. Polym. Environm.* 6 (1998) 197.

[52] Gattin R., Copinet A., Bertrand C., Couturier Y.: Biodegradation study of a starch and poly(lactic acid) co-extruded material in liquid, composting and inert mineral media, *Int. Biodeter. Biodegr.* 50 (2002) 25.

[53] Degli-Innocenti F., Tosin M., Bastioli C.: Evaluation of the biodegradation of starch and cellulose under controlled composting conditions, *J. Polym. Environm.* 6 (1998) 197.

[54] Pérez J., Muñoz-Dorado J., de la Rubia T., Martinez J.: Biodegradation and biological treatments of cellulose, hemicellulose and lignin: an overview, *Int. Microbiol.* 5 (2002) 53.

[55] Edgar K.J., Buchanan C.M., Debenham J.S., Rundquist P.A., Seiler B.D., Shelton M.C., Tindall D.: Advances in cellulose ester performance and application, *Prog. Polym. Sci.* 26 (2001) 1605.

[56] Gardner R.M., Buchanan C.M., Komarek R., Dorschel D., Bogus C., White A.W.: Compostability of cellulose acetate films, *J. Appl. Polym. Sci.* 52 (1994) 1477.

[57] Gu J.-D., Yang S., Welton R., Eberiel D., McCarthy S.P., Gross R.A.: Effect of environmental parameters on the degradability of polymer films in laboratory-scale composting reactors, *J. Environm. Polymer Degrad.* 2 (1994) 129.

[58] Glasser W.G., McCartney B., Samaranayake G.: Cellulose derivatives with low degree of substitution. 3. The biodegradability of cellulose esters using a simple enzyme assay, *Biotechnol. Prog.* 10 (1994) 214.

[59] Buchanan C.M., Dorschel D., Gardner R.M., Komarek R.J., Matosky A.J., White A.W., Wood M.D.: The influence of degree of substitution on blend miscibility and biodegradation of cellulose acetate blends, *J. Envir. Polym. Degrad.* 4 (1996).

[60] Samios E., Dart R.K., Dawkins J.V.: Preparation, characterization and biodegradation studies on cellulose acetates with varying degrees of substitution, *Polymer* 38 (1997) 3045.

[61] Frisoni G., Baiardo M., Scandola M.: Natural cellulose fibers: heterogenous acetylation kinetics and biodegradation behaviour, *Biomacromolecules* 2 (2001) 476.

[62] Ikejima T., Inoue Y.: Crystallization behaviour and environmental biodegradability of the blend films of poly(3-hydroxybutyric acid) with chitin and chitosan, *Carbohydr. Polymers* 41 (2000) 351.

[63] Lodha P., Netravali A.N.: Effect of soy protein isolate resin modifications on their biodegradation in a compost medium, *Polym. Deg. Stab.* 87 (2005) 465.

[64] Domenek S., Feuilloley P., Gratraud J., Morel M.-H., Guilbert S.: Biodegradability of wheat gluten based plastics, *Chemosphere* 54 (2004) 551.

[65] Abrusci C., Marquina D., Santos A., Del Amo A., Corrales T., Catalina F.: A chemiluminescence study on degradation of gelatine. Biodegradation by bacteria and fungi isolated from cinematographic films, *J. Photochem. Photobiol. A: Chemistry* 185 (2007) 188.

[66] Chiellini E., Cinelli P., Corti A., Kenawy E.R.: Composite films based on waste gelatin: thermal-mechanical properties and biodegradation testing, *Polym. Deg. Stab.* 73 (2001) 549.

[67] Fujimaki T.: Processability and properties of aliphatic polyesters, "BIONOLLE", synthesised by polycondensation reaction, *Polym. Deg. Stab.* 59 (1998) 209.

[68] Tserki V., Matzinos P., Pavlidou E., Vachliotis D., Panayiotou C.: Biodegradable aliphatic polyesters. Part I. Properties and biodegradation of poly(butylene succinate-co-butylene adipate), *Polym. Deg. Stab.* 91 (2006) 367.

[69] Tserki V., Matzinos P., Pavlidou E., Vachliotis D., Panayiotou C.: Biodegradable aliphatic polyesters. Part II. Synthesis and characterization of chain extended poly(butylene succinate-co-butylene adipate), *Polym. Deg. Stab.* 91 (2006) 377.

[70] Kim M.-N., Kim K.-H., Jin H.-J., Park J.-K., Yoon J.-S.: Biodegradability of ethyl and n-octyl branched poly(ethylene adipate) and poly(butylene succinate), *Eur. Polym. J.* 37 (2001) 1843.

[71] Bikiaris D.N., Papageorgiou G.Z., Achilias D.S.: Synthesis and comparative biodegradability studies of three poly(alkylene succinate)s, *Polym. Deg. Stab.* 91 (2006) 31.

[72] Marten E., Müller R.-J., Deckwer W.-D.: Studies on the enzymatic hydrolysis of polyesters. I. Low molecular mass model esters and aliphatic polyesters, *Polym. Deg. Stab.* 80 (2003) 485.

[73] Ahn B.D., Kim S.H., Kim Y.H., Yang J.S.: Synthesis and characterization of the biodegradable copolymers from succinic acid, and adipic acid with 1,4-butanediol, *J. Appl. Polym. Sci.* 82 (2001) 2808.

[74] Rizarrelli P., Puglisi C., Montaudo G.: Soil burial and enzymatic degradation in solution of aliphatic co-polyesters, *Polym. Deg. Stab.* 85 (2004) 855.

[75] Zhao J.-H., Wang X.-Q., Zeng J., Yang G., Shi F.-H., Yan Q.: Biodegradation of poly(butylene succinate-co-butylene adipate) by *Aspergillus versicolor, Polym. Deg. Stab.* 90 (2005) 173.

[76] Zhao J.-H., Wang X.-Q., Zeng J., Yang G., Shi F.-H., Yan Q.: Biodegradation of poly(butylene succinate) in compost, *J. Appl. Polym. Sci.* 97 (2005) 2273.

[77] Kim M.-N., Kim K.-H., Jin H.-J., Park J.-K., Yoon J.-S.: Biodegradability of ethyl and n-octyl branched poly(ethylene adipate) and poly(butylene succinate), *Europ. Polym. J.* 37 (2001) 1843.

[78] Jin H.-J., Kim D.-S., Lee B.-Y., Kim M.-N., Lee I.-M., Lee H.-S., Yoon J.-S.: Chain extension and biodegradation of poly(butylene succinate) with maleic acid units, *J. Polym. Sci.* 38 (2000) 2240.

[79] Pranamuda H., Tokiwa Y., Tanaka H.: Microbial degradation of an aliphatic polyester with a high melting point, poly(tetramethylene succinate), *Appl. Environ. Microbiol.* 61 (1995) 1828.

[80] Hayase N., Yano H., Kudoh E., Tsutsumi Ch., Ushio K., Miyahara Y., Tanaka S., Nakagawa K.: Isolation and characterization of poly(butylene succinate-*co*-butylene adipate)-degrading microorganism, *J. Biosci. Bioeng.* 97 (2004) 131.

[81] Tsutsumi C., Hayase N., Nakagawa K., Tanaka S., Miyahara Y.: The enzymatic degradation of commercial biodegradable polymers by some lipases and chemical degradation of them, *Macromol. Symp.* 197 (2003) 431.

[82] Scherer T.M., Clinton Fuller R., Lenz R.W., Goodwin S.: Hydrolase activity of an extracellular depolymerase from *Aspergillus fumigatus* with bacterial and synthethic polyesters, *Polym. Deg. Stab.* 64 (1999) 267.

[83] Jarerat A., Tokiwa Y.: Degradation of poly(tetramethylene succinate) by thermophilic actinomycetes, *Biotechnol. Lett.* 23 (2001) 647.

[84] Chandure A.S., Umare S.S.: Synthesis, characterization and biodegradation of low molecular weight polyesters, *Int. J. Polym. Mater.* 56 (2007) 339.

[85] Shirahama H., Kawaguchi Y., Aludin M.S., Yasuda H.: Synthesis and enzymatic degradation of high molecular weight aliphatic polyesters, J. Appl. Polym. Sci. 80 (2001) 340.

[86] Diamond M.J., Freedman B., Garibaldi J.A.: Biodegradable polyester films, *Int. Biodeter. Biodegrad.* 48 (2001) 219.

[87] Nikolic M.S., Djonlagic J.: Synthesis and characterization of biodegradable poly(butylene succinate-*co*-butylene adipate), *Polym. Deg. Stab.* 74 (2001) 263.

[88] Maeda H., Yamagata Y., Abe K., Hasegawa F., Machida M., Ishioka R., Gomi K., Nakajima T.: Purification and characterization of a biodegradable plastic-degrading enzyme from *Aspergillus oryzae, Appl. Microbiol. Biotechnol.* 67 (2005) 778.

[89] Müller R.-J., Kleeberg I., Deckwer W.-D.: Biodegradation of polyesters containing aromatic constituents, *J. Biotechnol.* 86 (2001) 87.

[90] Witt U., Müller R.-J., Deckwer W.-D.: Biodegradation of polyester copolymers containing aromatic compounds, *J. Macromol. Sci. – Pure Appl. Chem.* A32 (1995) 851.

[91] Müller R.-J., Witt U., Rantze E., Deckwer W.-D.: Architecture of biodegradable copolyesters containing aromatic constituents, *Polym. Deg. Stab.* 59 (1998) 203.

[92] Kleeberg I., Hetz C., Kroppenstedt R.M., Müller R.-J., Deckwer W.-D.: Biodegradation of aliphatic-aromatic copolyesters by *Thermonospora fusca* and other thermophilic compost isolates, *Appl. Environ. Microbiol.* 64 (1998) 1731.

[93] Kang H.J., Park S.S.: Characterization and biodegradability of poly(butylene adipate-*co*-succinate)/poly(butylene terephthalate) copolyester, *J. Appl. Polym. Sci.* 72 (1999) 593.

[94] Marten E., Müller R.-J., Deckwer W.-D.: Studies on the enzymatic hydrolysis of polyesters. II. Aliphatic-aromatic copolyesters, *Polym. Deg. Stab.* 88 (2005) 371.

[95] Müller R.-J., Schrader H., Profe J., Dresler K., Deckwer W.-D.: Enzymatic degradation of poly(ethylene terephthalate): rapid hydrolyse using a hydrolase from *T. fusca, Macromol. Rapid Commun.* 26 (2005) 1400.

[96] Mueller R.-J.: Biological degradation of synthetic polyesters – enzymes as potential catalysts for polyester recycling, *Proc. Biochem.* 41 (2006) 2124.

[97] Maeda Y., Maeda T., Yamaguchi K., Kubota S., Nakayama A., Kawasaki N., Yamamoto N., Aiba S.: Synthesis and characterization of novel biodegradable copolyesters by transreaction of

poly(ethylene terephthalate) with copoly(succinic anhydride/ethylene oxide), *J. Polym. Sci. Part A: Polymer Chemistry* 38 (2000) 4478.

[98]   Edlund U., Albertsson A.-Ch.: Degradable polymer microspheres for controlled drug delivery, *Adv. Polym. Sci.* 157 (2002) 67.

[99]   Krasowska K., Heimowska A., Rutkowska M.: Enzymatic and hydrolytic degradation of poly (ε-caprolactone), *Intern. Polymer Sci. Technol.* 33 (2006) T/57.

[100] Eldsäter C., Erlandsson B., Renstad R., Albertsson A.-C., Karlsson S.: The biodegradation of amorphous and crystalline regions in film-blown poly(ε-caprolactone), *Polymer* 41 (2000) 1297.

[101] Mezzanote V., Bertani R., Degli Innocenti F., Tosin M.: Influence of inocula on the results of bio-degradation tests, *Polym. Deg. Stab.* 87 (2005) 51.

[102] Nakasaki K., Ohtaki, Takano H.: Biodegradable plastic reduces ammonia emission during com-posting, *Polym. Deg. Stab.* 70 (2000) 185.

[103] Ohtaki A., Sato N., Nakasaki K.: Biodegradation of poly(ε-caprolactone) under controlled com-posting conditions, *Polym. Deg. Stab.* 61 (1998) 449.

[104] Ohtaki A., Akakura N.., Nakasaki K.: Effects of temperature and inoculum on the degradability of poly(ε-caprolactone) during composting, *Polym. Deg. Stab.* 62 (1998) 279.

[105] Gattin R., Cretu A., Barbier-Baudry D.: Effect of the remaining lanthanide catalysts on the hydro-lytic and enzymatic degradation of poly(ε-caprolactone), *Macromol. Symp.* 197 (2003) 455.

[106] Ohkawa K., Kim H., Lee K.: Biodegradation of electrospun poly(ε-caprolactone) non-woven fab-rics by pure-cultured soil filamentous fungi, *J. Polym. Environ.* 12 (2004) 211.

[107] Funabashi M., Ninomiya F., Kunioka M.: Biodegradation of polycaprolactone powders proposed as reference test materials for international standard of biodegradation evaluation method, *J. Polym. Environ.* 15 (2007) 7.

[108] Botines E., Rodríguez-Galán A., Puiggalí J.: Poly(ester amide)s derived from 1,4-butanediol, adipic acid and 1,6-aminohexanoic acid: characterization and degradation studies, *Polymer* 43 (2002) 6073.

[109] Ferré T., Franco L., Rodríguez-Galán A., Puiggalí J.: Poly(ester amide)s derived from 1,4-butane-diol, adipic acid and 1,6-aminohexanoic acid. Part II: composition changes and fillers, *Polymer* 44 (2003) 6139.

[110] Lozano M., Franco L., Rodríguez-Galán A., Puiggalí J.: Poly(ester amide)s derived from 1,4-butanediol, adipic acid and 1,6-aminohexanoic acid. Part III: substitution of adipic acid units by terephthalic acid units, *Polym. Deg. Stab.* 85 (2004) 595.

[111] Chen X., Gonsalves K.E., Cameron J.A.: Further studies on biodegradation of aliphatic poly(ester-amides), *J. Appl. Polym. Sci.* 50 (1993) 1999.

[112] Rodríguez-Galán A., Fuentes L., Puiggalí J.: Studies on the degradability of a poly(ester amide) derived from L-alanine, 1,12-dodecanediol and 1,12-dodecanedioic acid, *Polymer* 41 (2000) 5967.

[113] Grigat E, Koch R., Timmermann R.: BAK 1095 and BAK 2195: completely biodegradable syn-thetic thermoplastics, *Polym. Deg. Stab.* 59 (1998) 223–226

[114] Jayeskara R., Sheridan S., Lourbakos E., Beh H., Christie G.B.Y., Jenkins M., Halley P.B., McGlashan S., Lonergan G.T.: Biodegradation and ecotoxicity evaluation of a Bionolle and starch blend and its degradation products in compost, *Ind. Biodeter. & Biodeg.* 51 (2003) 77–81.

[115] Park C., Kim E.Y., Yoo Y.T., Im S.S.: Effect of hydrophilicity on the biodegradability of polyes-teramides, *J. Appl. Polym. Sci.* 90 (2003) 2708.

[116] Chiellini E., Corti A., Solaro R.: Biodegradation of poly(vinyl alcohol) based blown films under different environmental conditions, *Polym. Deg. Stab.* 64 (1999) 305.

[117] Corti A., Solaro R., Chiellini E.: Biodegradation of poly(vinyl alcohol) in selected mixed culture and relevant culture filtrate, *Polym. Deg. Stab.* 75 (2002) 447.

[118] Chiellini E., Corti A., D'Antone S., Solaro R.: Biodegradation of poly(vinyl alcohol) based mate-rials, *Prog. Polym. Sci.* 28 (2003) 963.

[119] Hoffmann J., Řezníčková, I., Kozáková J., Růžička J., Alexy P., Bakoš D., Precnerová L.: Assessing biodegradability of plastics based on poly(vinyl alcohol) and protein wastes, *Polym. Deg. Stab.* 79 (2003) 511.

[120] David C., De Kesel C., Lefebre F., Weiland M.: The biodegradation of polymers: a recent studies, *Angew. Makromol. Chem.* 216 (1994) 21.

[121] Chiellini E., Corti A., Del Sarto G., D'Antone S.: Oxo-biodegradable polymers – effect of hydrolysis degree on biodegradation of poly(vinyl alcohol), *Polymer Deg. Stab.* 91 (2006) 3397.

[122] Matsumura S., Tomizawa N., Toki A., Nishikawa K., Toshima K.: Novel poly(vinyl alcohol)-degrading enzyme and the degradation mechanism, *Macromolecules* 32 (1999) 7753.

[123] Watanabe M., Kawai F.: Numerical simulation for enzymatic degradation of poly(vinyl alcohol), *Polym. Deg. Stab.* 81 (2003) 393.

[124] Solaro R., Corti A., Chiellini E.: Biodegradation of poly(vinyl alcohol) with different molecular weights and degree of hydrolysis, *Polym. Advanc. Technol.* 11 (2000) 873.

[125] Chiellini E., Corti A., Politi B., Solaro R.: Adsorption/desorption of polyvinyl alcohol on solid substrates and relevant biodegradation, *J. Polym. Environ.* 8 (2000) 67.

[126] Ohkita T., Lee S.-H.: Thermal degradation and biodegradability of poly(lactic acid)/corn starch biocomposites, *J. Appl. Polym. Sci.* 100 (2006) 3009.

[127] Tsuji H., Kidokoro Y., Mochizuki M.: Enzymatic degradation of biodegradable polyester composites of poly(L-lactic acid) and poly(ε-caprolactone), *Macromol. Mater. Eng.* 291 (2006) 1245.

[128] Imam S.H., Chen L., Gordon S.H., Shogren R.L., Weisleder D., Greene R.V.: Biodegradation of injection molded starch-poly (3-hydroxybutyrate-*co*-3-hydroxyvalerate) blends in a natural compost environment, *J. Envionm. Polym. Deg.* 6 (1998) 91.

[129] Rychter P., Biczak R., Herman B., Smyłła A., Kurcok P., Adamus G., Kowalczuk M.: Environmental degradation of polyester blends containing atactic poly(3-hydroxybutyrate). Biodegradation in soil and ecotoxicological impact, *Biomacromolecules* 7 (2006) 3125.

[130] Dubois P., Narayan R.: Biodegradable compositions by reactive processing of aliphatic polyester/polysaccharide blends, *Macromol.Symp.* 198 (2003) 233.

[131] Singh R.P., Pandey J.K., Rutot D., Degée Ph., Dubois Ph.: Biodegradation of poly(ε-caprolactone)/starch blends and composites in composting and culture environments: the effect of compatibilization on the inherent biodegradability of the host polymer, *Carbohydr. Res.* 338 (2003) 1759.

[132] Dos Santos Rosa D., Rodrigues T.C., das Graças Fassina Guedes C., Calil M.R.: Effect of thermal aging on the biodegradation of PCL, PHB-V, and their blends with starch in soil compost, *J. Appl. Polym. Sci.* 89 (2003) 3539.

[133] Nugroho P., Mitomo H., Yoshi F., Kume T., Nishimura K.: Improvement of processability of PCL and PBS blend by irradiation and its biodegradability, *Macromol. Mater. Eng.* 286 (2001) 316.

[134] Wu C.-S.: A comparison of the structure, thermal properties, and biodegradability of polycaprolactone/chitosan and acrylic acid grafted polycaprolactone/chitosan, *Polymer* 46 (2005) 147.

[135] Rohindra D., Sharma P., Khurma J.: Soil and microbial degradation study of poly(ε-caprolactone)–poly(vinyl butyral) blends, *Macromol. Symp.* 224 (2005) 323.

[136] Dos Santos Rosa D., Volponi J.E., das Graças Fassina Guedes C.: Biodegradation and the dynamic mechanical properties of starch gelatinization in poly(ε-caprolactone)/corn starch blends, *J. Appl. Polym. Sci.* 102 (2006) 825.

[137] Tsuji H., Horikawa G.: Porous biodegradable polyester blends of poly(L-lactic acid) and poly(ε-caprolactone): physical properties, morphology, and biodegradation, *Poly. Int.* 56 (2007) 258.

[138] Kim C.-H., Jung K.-M., Kim J.-S., Park J.-K.: Modification of aliphatic polyesters and their reactive blends with starch, *J. Polym. Environ.* 12 (2004) 179.

[139] Tsuji H., Horikawa G., Itsuno S.: Melt-processed biodegradable polyester blends of poly(L-lactic acid) and poly(ε-caprolactone): effects of processing conditions on biodegradation, *J. Appl. Polym. Sci.* 104 (2007) 831.

[140] Mariani P.D.S.C., Vinagre Neto A.P., da Silva Jr J.P., Cardoso E.J.B.N., Esposito E., Innocentini-Mei L.H.: Mineralization of poly(ε-caprolactone)/adipate modified starch blend in agricultural soil, *J. Polym. Environ.* 15 (2007) 19.

[141] Zhang P., Huang F., Wang B.: Characterization of biodegradable aliphatic/aromatic copolyesters and their starch blends, *Polym.-Plast. Technol. Eng.* 41 (2002) 273.

[142]   Rutkowska M., Krasowska K., Heimowska A., Kowalczuk M.: Degradation of the blends of natural and synthetic copolyesters in different natural environment, *Macromol. Symp.* 197 (2003) 421.

[143]   Ratto J.A., Stenhouse P.J., Auerbach M., Mitchell J., Farrell R.: Processing, performance and biodegradability of a thermoplastic aliphatic polyester/starch system, *Polymer* 40 (1999) 6777.

[144]   Ikejima T., Cao A., Yoshie N., Inoue Y.: Surface composition and biodegradability of poly(3-hydroxybutyric acid)/poly(vinyl alcohol) blend films, *Polym. Deg. Stab.* 62 (1998) 463.

[145]   Alexy P., Bakoš D., Hanzelová S., Kukolíková L., Kupec J., Charvátová K., Chiellini E., Cinelli P.: Poly(vinyl alcohol)–collagen hydrolysate thermoplastic blends: I. Experimental design optimization and biodegradation behaviour, *Polymer Testing* 22 (2003) 801.

[146]   Alexy P, Bakoš D., Hanzelová S., Crkoňová G., Kramárová Z., Hoffman J., Julinová M., Chiellini E., Cinelli P.: Poly(vinyl alcohol)–collagen hydrolysate thermoplastic blends: II. Water penetration and biodegradability of melt extruded films, *Polymer Testing* 22 (2003) 811.

[147]   Takasu A., Aoi K., Tsuchiya M., Okada M.: New chitin-based polymer hybrids. 4: Soil burial degradation behaviour of poly(vinyl alcohol)/chitin derivative miscible blends, *J. Appl. Polym. Sci.* 73 (1999) 1171.

[148]   Tsuji H., Muramatsu H.: Blends of aliphatic polyesters: V. Non-enzymatic and enzymatic hydrolysis of blends from hydrophobic poly(L-lactide) and hydrophilic poly(vinyl alcohol), *Polym. Deg. Stab.* 71 (2001) 403.

[149]   Guohua Z., Ya L., Cuilan F., Min Z., Caiqiong Z., Zongdao C.: Water resistance, mechanical properties and biodegradability of methylated-corn starch/poly(vinyl alcohol) blend film, *Polym. Deg. Stab.* 91 (2006) 703.

[150]   Corti A., Cinelli P., D'Antone S., Kenawy E.-R., Solaro R.: Biodegradation of poly(vinyl alcohol) in soil environment: influence of natural organic fillers and structural parameters, *Macromol. Chem. Phys.* 203 (2002) 1526.

[151]   Jayasekara R., Harding I., Bowater I., Christie G.B.Y., Lonergan G.T.: Biodegradation by composting of surface modified starch and PVA blended films, *J. Polym. Environ.* 11 (2003) 49.

[152]   Yang H.-S., Yoon J.-S., Kim M.-N.: Effects of storage of a mature compost on its potential for biodegradation of plastics, *Polym. Deg. Stab.* 84 (2004) 411.

[153]   Kim M.-N., Lee A.-R., Yoon J.-S., Chin I.-J.: Biodegradation of poly(3-hydroxybutyrate), Sky-Green and Mater-Bi by fungi isolated from soils, *Eur. Polym. J.* 36 (2000) 1677.

[154]   Tsuji H., Suzuyoshi K., Tezuka Y., Ishida T.: Environmental degradation of biodegradable polyesters: 3. Effects of alkali treatment on biodegradation of poly(ε-caprolactone) and poly((R) -3-hydroxybutyrate) films in controlled soil, *J. Polym. Environ.* 11 (2003) 57.

[155]   Ishigaki T., Sugano W., Nakanishi A., Tateda M., Ike M., Fujita M.: The degradability of biodegradable plastics in aerobic and anaerobic waste landfill model reactors, *Chemosphere* 54 (2004) 225.

[156]   Massardier-Nageotte V., Pestre C., Cruard-Pradet T., Bayard R.: Aerobic and anaerobic biodegradability of polymer films and physico-chemical characterization, *Polym. Deg. Stab.* 91 (2006) 620.

[157]   Renstad R., Karlsson S., Sandgren A., Albertsson A.-Ch.: Influence of processing additives on the degradation of melt-processing films of poly(ε-caprolactone) and poly(lactic acid), *J. Envir. Polym. Degrad.* 6 (1998) 209.

[158]   Calil M.R., Gaboardi F., Guedes C.G.F., Rosa D.S.: Comparison of the biodegradation of poly(ε-caprolactone) (PCL), cellulose acetate (CA) and their blends by the Sturm test and selected fungi, *Polymer Testing* 25 (2006) 597.

[159]   Avérous L.: Biodegradable multiphase systems based on plasticized starch: a review, *Polymer Reviews* 44 (2004) 231.

Chapter 7
Ecotoxicological assessment

Chapter 7
Ecotoxicological assessment

7.1. INTRODUCTION

Composting is the most relevant waste treatment technology for biodegradable plastics. Technologies such as composting used for the disposal of food and yard waste, accounting for 25–40% of the municipal solid waste, are the most suitable for the disposal of biodegradable materials together with soiled or food-containing paper [1]. The advantages of composting compared to other waste treatment techniques are not only the relatively low cost but also technical reasons. However, the influence of polymeric material on the composting process should be minimal. Thus, for the acceptance of compostable polymers it is important to develop appropriate testing methods and standards.

The European Committee for Standardization (CEN) has developed a standard with requirements for compostable products – EN 13432 [2]. With regard to acceptance for composting, basically four characteristics are required:

1. Biodegradability
2. Disintegration
3. Effect on the biological treatment process
4. Effect on the quality of the resulting compost

The material characteristics include a minimum content of organic matter determined as volatile solids (minimum 50%) and a maximum level of heavy metals. To demonstrate biodegradability, it is possible to use several internationally accepted standard methods for determining the biodegradability of organic compounds. The controlled aerobic composting test ISO 14855 [3] is recommended as a suitable test for the aerobic degradation of polymers. Two aquatic tests are also listed, the respirometric test ISO 14851 [4] and the CO_2 evolution test ISO 14852 [5]. If the biodegradability of a packaging material has been shown with laboratory tests, investigations are required to confirm the disintegration of a packaging material in the form in which it will be later used, e.g. as film of a certain thickness or shaped articles. Disintegration can be tested in a pilot-scale composting plant or in a real full-scale treatment facility. Possible negative effects on compost quality of adding the compostable products to the biowaste shall be determined by direct comparison with the quality of compost obtained from a control composting experiment in which no product was added to the biowaste. The evaluation criteria include the density, total dry solids, volatile solids, salt content, pH and ecotoxicity of the final compost. With regard to the evaluation of ecotoxicity, the standard requires a plant growth test on the final compost using two higher plants.

In the USA the ASTM D 6400 standard [6] identifies three key criteria for materials and products to be compostable. A compostable plastic refers to a plastic that breaks down in a composting environment as fast as the surrounding material and leaves no visible, distinguishable or toxic material. In order for a plastic to meet this standard it must biodegrade, disintegrate and be safe for the environment – produce no harmful by-products or hinder the ability of the compost to support plant growth. Ecotoxicity is determined by measuring plant growth and germination.

It must be noted that the ecotoxicity tests were originally designed for the testing and evaluation of pure chemicals [7, 8]. Ecotoxicity tests based on the OECD Guidelines for Testing of Chemicals (numbers 201, 202, 207 and 208) are usually employed. [7].

7.2. DEFINITIONS

7.2.1 Ecotoxicology

According to IUPAC ecotoxicology is the study of the toxic effects of chemical and physical agents on all living organisms, especially on populations and communities within defined ecosystems. Ecosystems are defined as the grouping of organisms (microorganisms, plants, animals) interacting together, with and through their physical and chemical environments, to form a functional entity. Ecotoxicologists seek to predict the impacts of chemicals on ecosystems.

7.2.2 Ecotoxicity

According to EPA (US Environmental Protection Agency) ecotoxicity means toxic (harm from chemicals) effects on plants and animals, populations, or communities.

The goal of ecotoxicity is to understand the concentration of chemicals at which organisms in the environment will be affected.

With regard to compostable polymer materials ecotoxicity refers to the potential environmental toxicity of residues, leachate, or volatile gases produced by the plastics during biodegradation or composting [9].

Ecotoxicity tests are used to determine the toxicity of a substance to one or more species. A ecotoxicity test measures the degree of response produced by exposure to a specific level of substance (i.e. concentration or dose) compared to an unexposed control. Standard ecotoxicity tests measure important physiological and ecologically relevant responses such as lethality, growth and reproduction. Results are expressed as LD, LC, ED, EC or NOEC:

- **LD or LC – lethal dose or lethal concentration:** The dose or concentration that produces a specified level of mortality in the test population within a specified time, e.g. LC_{50} is the median lethal concentration or the concentration of a substance at which 50% of the test population are killed.
- **ED or EC – effective dose or effective concentration:** The dose or concentration that produces a specified level of effect in the test population within a specified time, e.g. EC_{50} is the median effective concentration or the concentration of a substance at which 50% of the test population are affected.
- **NOEC – no observed effect concentration:** The highest level of substance in an ecotoxicity test that did not cause harmful effects in a tested organism, e.g. in a plant or animal.
- **LOEC – lowest observed effect concentration.**
- **Phytotoxicity:** Toxicity to plants.

7.3. METHODS

Ecotoxicity tests have been developed for the risk assessment of water pollution and contaminated soils. The main goal for ecotoxicological assessment of compostable polymers is to ensure that no harmful substances are released in the environment during their degradation and after they are degraded.

The possible ecotoxicity tests for compost applications are discussed by Kapanen and Itävaara [10].

The methods of the evaluation of the ecotoxicity of compostable polymer materials are mainly based on the use of:

- Plants
- Soil fauna (earthworms)
- Aquatic fauna (*Daphnia*)
- Algae (green algae)
- Microbes (luminescent bacteria)

The main ecotoxicity tests and standards used for compostable polymers are summarized in Tables 7.1 and 7.2.

7.3.1 Plant phytotoxicity testing

Phytotoxicity is expressed as a delay of seed germination, inhibition of plant growth or any adverse effect on plants caused by specific substances (phytotoxines) or growing conditions.

While a product may not negatively impact plant growth in the short term, over time it could become phytotoxic due to the build-up of inorganic materials, which could potentially lead to a

Table 7.1. Ecotoxicity tests

Test species	Parameters measured	Results expressed[*]	Standards
Plants	<ul><li>fresh (and/or dry weight) of the shoot</li><li>the number of seedlings emerging</li><li>the number of plants remaining at harvest</li><li>the germination rate</li></ul>	<ul><li>number and % emergence as compared to the controls</li><li>biomass measurements (shoot weight or shoot height as % of the controls)</li><li>% visual injury</li></ul>	OECD 208 ISO 11269-1:1993 ISO 11269-2:1995 ASTM E 1598-94
Earthworms	Mortality of earthworms	LC 50, EC50, NOEC	OECD 207 ISO 11268-1 ASTM E 1676-04
Daphnia	The effects on the swimming capability of *Daphnia* during 24 hours' exposure	24 h EC 50	OECD 202 ISO 6341:1996 EN ISO 6341:1996 DIN 38412 part 30
Algae	The effects of a substance on the growth of a unicellular green algal species	EC values NOEC	OECD 201 ISO 8692:2004 EN ISO 8692:2004 ASTM E 1218-04
Luminescent bacteria *Vibrio fisheri*	Inhibition of the natural light emission of the microorganisms	% inhibition, G_L, EC_{20} or EC_{50}	ISO 11348: 1998 EN ISO 11348: 1998 DIN 38412 part 34

[*] Definitions explained in Table 7.5.

Table 7.2. Standards used for ecotoxicity tests

Number	Title	Remarks
OECD 201	Alga, Growth Inhibition Test (Updated Guideline, adopted 7 June 1984)	OECD Guidelines for Testing of Chemicals
OECD 202	*Daphnia* sp. Acute Immobilization Test (Updated Guideline, adopted 13April 2004)	OECD Guidelines for Testing of Chemicals
OECD 207	Earthworm, Acute Toxicity Tests (Original Guideline, adopted 4 April 1984)	OECD Guidelines for Testing of Chemicals
OECD 208	Terrestrial Plants, Growth Test (Original Guideline, adopted 4 April 1984)	OECD Guidelines for Testing of Chemicals
		Recommended by EN 13 432
ISO 11269-1: 1993	Determination of the effects of pollutants on soil flora. 1. Method for the measurement of inhibition of root growth	
ISO 11269-2:1995	Determination of the effects of pollutants on soil flora. 2. Effects of chemicals on the emergence and growth of higher plants	
ISO 6341:1996	Water quality. Determination of the inhibition of the mobility of *Daphnia magna* Straus (Cladocera, Crustacea). Acute toxicity test	Equivalent EN 6341:1996
ISO 8692:2004	Water quality. Freshwater algal growth inhibition test with unicellular green algae	Equivalent EN ISO 8692:2004
ISO 11348-1: 1998	Water quality. Determination of the inhibitory effect of water samples on the light emission of *Vibrio fisheri* (luminescent bacteria test) – Part 1: Method using freshly prepared bacteria	
ISO 11348-2: 1998	Water quality. Determination of the inhibitory effect of water samples on the light emission of *Vibrio fisheri* (luminescent bacteria test) – Part 2: Method using liquid-dried bacteria	
ISO 11348-3: 1998	Water quality. Determination of the inhibitory effect of water samples on the light emission of *Vibrio fisheri* (luminescent bacteria test) – Part 3: Method using freeze-dried bacteria	
ASTM E 1676-04	Standard guide for conducting a laboratory soil toxicity test or bioaccumulation tests with the lumbricid earthworm *Eisenia foetida* and the enchytraeid potworm *Enchytraeus albidus*	
ASTM E 1218-04	Standard guide for conducting static toxicity tests with microalgae	
ASTM E 1598-94	Standard practice for conducting early seedling growth tests	
ISO 11269-1:1993	Soil quality. Determination of the effects of pollutants on soil flora – method for the measurement of inhibition of root growth (1993)	

Table 7.2. (*Continued*)

Number	Title	Remarks
ISO 11269-2:1995	Soil quality. Determination of the effects of pollutants on soil flora – effects of chemicals on the mergence and growth of higher plants	
ISO 11268-1:1993	Soil quality. Effects of pollutants on earthworms (*Eisenia fetida*) – Part 1: Determination of acute toxicity using artificial soil substrate	

reduction in soil productivity. For this reason some manufacturers use plant phytotoxicity testing on the finished compost that contains degraded polymers.

Phytotoxicity testing can be conducted on two classes of flowering plants. These are monocots (plants with one seed leaf) and dicots (plants with two seed leaves). Representatives from both of these classes are typically used in toxicity testing – summer barley to represent monocots and cress to represent dicots. Test involves measuring the yield of both of these plants obtained from the test compost and from the control compost.

In the OECD 208 test method, recommended by the EN 13432 standard, the test substance is incorporated at various concentrations into soil in which the seeds are sown. The number of seedlings that emerge is recorded. At least two weeks after 50% of the seedlings have emerged in the control, the plants are harvested and weighed. A minimum of three species should be selected for testing, at least one from each of the categories listed in Table 7.3. The results are expressed as the effect of the test substance on emergence (EC50) and the effect on growth (LC50) (see Table 7.5).

Table 7.3. Plant species recommended by OECD Guideline 208

Category	Group	Family	Common name	Scientific name
1	Monocotyledonae	Poaceae	ryegrass	*Lolium perenne*
1	Monocotyledonae	Poaceae	rice	*Oryza sativa*
1	Monocotyledonae	Poaceae	oat	*Avena sativa*
1	Monocotyledonae	Poaceae	wheat	*Triticum aestivum*
1	Monocotyledonae	Poaceae	sorghum	*Sorghum bicolor*
2	Dicotyledonae	Brassicaceae	mustard	*Brassica alba*
2	Dicotyledonae	Brassicaceae	rape	*Brassica napus*
2	Dicotyledonae	Brassicaceae	radish	*Raphanus sativus*
2	Dicotyledonae	Brassicaceae	turnip	*Brassica rapa*
2	Dicotyledonae	Brassicaceae	Chinese cabbage	*Brassica campestris* var. *chinesis*
3	Dicotyledonae	Fabaceae	vetch	*Vicia sativa*
3	Dicotyledonae	Fabaceae	mung bean	*Phaseolus aureus*
3	Dicotyledonae	Fabaceae	red clover	*Trifolium pratense*
3	Dicotyledonae	Fabaceae	fenugreek	*Trifolium ornithopodioides*
3	Dicotyledonae	Asteraceae	lettuce	*Lactuca sativa*
3	Dicotyledonae	Brassicaceae	cress	*Lepidium sativum*

Table 7.4. Plant species recommended by ISO 11269-2

Category	Group	Family	Common name	Scientific name
1	Monocotyledonae	Poaceae	barley (spring or winter)	*Hordeum vulgare* L.
1	Monocotyledonae	Poaceae	Rye	*Secale cerale* L.
1	Monocotyledonae	Poaceae	ryegrass, perennial	*Lolium perenne* L.
1	Monocotyledonae	Poaceae	rice	*Oryza sativa* L.
1	Monocotyledonae	Poaceae	oat (common or winter)	*Avena sativa* L.
1	Monocotyledonae	Poaceae	wheat, soft	*Triticum aestivum* L.
1	Monocotyledonae	Poaceae	sorghum, common (or shattercane or durra, white or millet, great)	*Sorghum bicolor* L.
2	Dicotyledonae	Brassicaceae	Chinese cabbage	*Brassica campestris* L. var. *Chinesis*
2	Dicotyledonae	Brassicaceae	cress, garden	*Lepidium sativum* L.
2	Dicotyledonae	Brassicaceae	mustard, white	*Sinapis alba*
2	Dicotyledonae	Brassicaceae	rape (or rape (summer) or rape (winter))	*Brassica napus* (L.) ssp. *napus*
2	Dicotyledonae	Brassicaceae	radish, wild	*Raphanus sativus* L.
2	Dicotyledonae	Brassicaceae	turnip, wild	*Brassica rapa* ssp. *rapa* (DC.) Metzg.
2	Dicotyledonae	Brassicaceae	bird's foot clover, fenugreek	*Trifolium ornithopodioides* L.
2	Dicotyledonae	Brassicaceae	lettuce	*Lactuca sativa* L.
2	Dicotyledonae	Brassicaceae	tomato	*Lycopersicon esculentum*
2	Dicotyledonae	Brassicaceae	bean	*Phaseolus aureus* Roxb.

7.3.2 Animal toxicity test

Animal testing is generally carried out using earthworms (as representative soil dwelling organisms) and *Daphnia* (as representative aquatic organisms) [8]. Earthworms are very sensitive to toxicants. Since earthworms feed on soil, they are suitable for testing the toxicity of compost.

In the acute toxicity test, earthworms are exposed to high concentrations of the test material for short periods of time. According to OECD guideline 207 earthworms are exposed to soil and compost in varying amounts. Following 14 days of exposure, the number of surviving earthworms is counted and weighed and the per cent survival rates are calculated. The earthworms are exposed to several ratios of compost and soil mixtures.

Compost worms (*Eisenia fetida*) are used for testing the toxicity of biodegradable plastic residues. These worms are very sensitive to metals such as tin, zinc, heavy metals and high acidity. The results are expressed as LC50 (see Table 7.5).

Daphnia (commonly named water flea) is one of the most common crustaceans to be found in lakes, ponds and streams. The *Daphnia* toxicity test can establish whether degradation products present in liquid pose any problem to surface water bodies. In the test, *Daphnia* are placed in test solutions for 24 hours. After exposure the number of surviving organisms is counted and the per cent mortality is calculated. The results are usually expressed as 24 h EC 50 (Table 7.5).

Table 7.5. Definitions related to ecotoxicity testing

Symbols	Definitions	Standards
EC50	The concentration at which the change in growth is 50% of that of the control. *Growth* is expressed in terms of plant weight	OECD 208 Plants
LC50	The concentration at which the change in emergence is 50% of that of the control. *Emergence* is the appearance of the seedling above the soil surface	OECD 208 Plants
LC50	The median lethal concentration, i.e. that concentration of the test substance which kills 50% of the test animals within the test period	OECD 207 Earthworms
24 h EC 50	The concentration estimated to immobilize 50% of the *Daphnia* after 24 h exposure. *Immobilization* – those animals not able to swim within 15 seconds after gentle agitation of the test container are considered to be immobile	OECD 202 Daphnia
EC 50	The concentration of test substance which results in a 50% reduction in either growth or growth rate relative to the control. *Growth* is the increase in cell concentration over the test period. *Growth rate* is the increase in cell concentration per unit of time	OECD 201 Algae
NOEC	No observed effect concentration (NOEC) is the highest concentration tested at which the measured parameter(s) show(s) no significant inhibition of growth relative to control values	OECD 201 Algae
% Inhibition	The percentage inhibition of the light emission compared with that of the control	ISO 11348-2 EN ISO 11348-2
EC_{20}	The concentration of a sample that causes exactly 20% inhibition	ISO 11348-2 EN ISO 11348-2
EC_{50}	The concentration of a sample that causes exactly 50% inhibition	ISO 11348-2 EN ISO 11348-2
G_L	First dilution level of a sample that causes less than 20% inhibition	DIN 38412 part 34

7.3.3 Algal test

Algae belong to a group of predominantly aquatic, photosynthetic organisms of the kingdom *Protista*. These very simple chlorophyl-containing organisms are important as primary producers of organic matter of the food chain. They also provide oxygen for aquatic life.

Freshwater algae are sensitive bioindicators. In the test the batch cultures of the unicellular algae are incubated with media and a water-soluble sample in a flask for many generations. After 72 h the growth inhibition compared with the control sample is measured. The results are expressed as EC50 and NOEC (Table 7.5).

7.3.4 Luminescent bacteria test

The bioluminescence produced by the marine bacterium *Vibrio fischeri* (formerly *Photobacterium phosphoreum*) is the basis for several toxicity bioassays that have been used to assess chemical toxicity of contaminated water, sediment and soil [11]. The test organism emits light as a result of normal functioning of its metabolic pathways [12]. Changes in bacterial health interrupt metabolic activity, therefore reducing light output. Reduction in light emission can be measured photometrically to provide a quantitative measure of toxicity.

Standardized test

Determination of acute toxicity using liquid-dried luminescent bacteria is based on the measurement of the natural light emission of these microorganisms. The inhibition of the light emission in the presence of the sample is determined against a non-toxic control.

The applicability of the test includes municipal and industrial wastewater, aqueous eluates from soil and waste, solution and surface water.

Results are reported as percentage inhibition and expressed as G_L, EC_{20} or EC_{50} (Table 7.5). The higher the percentage inhibition of the light emission, the more harmful is the sample's effect on the microorganisms.

Flash method

The test is based on measuring the luminescence of luminescent bacteria *Vibrio fisheri* and was developed for measuring the toxicity of solid and coloured samples [13].

In the method the signal from the sample is measured immediately after coming into contact with the sample and again after an incubation period of less than one minute.

Definitions related to ecotoxicity testing are summarized in Table 7.5.

7.4. COMPOSTABLE POLYMERS ECOTOXICITY TESTING

The potential toxicity of starch–polyester blends during biodegradation was tested against the earthworm *Eisenia foetida* [14] (see Table 7.6). As an indicator a change in weight and survival rate for earthworms exposed to the polymer directly and for those added after the polymer degraded was used. The significant differences in weight changes between the earthworms incorporated with the polymers and those exposed to the breakdown products were observed. The pretreatment of the starch polymer films by the compost leads to production of metabolites that are beneficial to the reproduction of earthworms. Regarding pathology no evidence of mortality was found. It was stated that this polymer is safe based on the criterion of number of juveniles and the duration of test.

Ecotoxicological assessment of different natural and synthetic compostable polymers was reported by Fritz *et al.* [15]. Based on the OECD and DIN standards for higher plants, earthworms, *Daphnia magna* and bioluminescent bacteria *Vibrio fisheri* the average ecotoxicity calculated from six individual biotest data was proposed. The average ecotoxicity was built by equal weight of each inhibition result. It was found that natural polymers (starch, cellulose and wood) inhibit the plant growth significantly as long as they were not fully degraded but increased the plant compatibility of the soil at a later time (after 160 days). The other materials tested (two biogenic materials and synthetic poly(esteramide)) inhibit during and after biodegradation.

The ecotoxicological impact of lactic acid-based polymers was evaluated by biotests, i.e. by the Flash test, measuring the inhibition of light production of *Vibrio fisheri*, and by plant growth tests with cress, radish, and barley [16]. Poly(lactic acid)s, poly(esterurethane)s and poly(esteramide)s of different structure units were biodegraded under controlled composting conditions and the quality of the compost was evaluated. Moreover, toxicity of polymer components, i.e. lactic acid, 1,4-butanediol, stannous octoate, 1,6-hexamethylene diisocyanate, 1,6-hexamethylenediamine, 1,4-butane diisocyanate, 1,4-butane diamine, lactide, succinic anhydride, succinic acid, and 2,2-bis(2-oxazoline) was assessed using the standardized ISO luminescent bacteria test and the Flash test. Both the Flash test and plant growth experiments revealed the toxic effect and its relation to 1,6-hexamethylene diisocyanate (HMDI) concentration in HMDI linked polymers. Poly(esterurethane), where lactic acid prepolymers were linked with 1,4-butane diisocyanate (BDI), did not exhibit any ecotoxicological effect, and neither did poly(esteramide) or poly(lactide). However, the results strongly suggest that 1,6-hexamethylene diisocyanate, which is frequently used in urethane chemistry and as a connecting agent in different polymers, should not be used as a structure unit in biodegradable polymers because of the environmental risk.

It was stated that the Flash test is fast and a valuable toxicity test for pre-evaluation of the quality of compost, and it has been found to work well in composting studies.

Lactic-based poly(esterurethanes) compost samples from a municipal waste composting plant and from a controlled composting test were studied [17]. Ecotoxicity of polymers linked with different chain extenders (HMDI or BDI) was evaluated using the Flash test and a plant growth assay. It was found that both biotests showed toxicity in relation to the amount of HMDI in the polymer. None of the other polymers tested had any toxic response. The effect of the matrix where biodegradation occurs on the applicability of the biotests was also studied. It was observed that the degree of maturity of the compost had an effect on the light production by *Vibrio fisheri* and on the growth of radish. For example, immature compost, i.e. composted for less than three months, was toxic in both biotests.

Standardized DIN tests using *Daphnia magna* and luminescent bacteria were applied to assess the ecotoxicity of synthetic aliphatic–aromatic copolyester [18]. No significant toxicological effect was observed, neither for the monomeric intermediates nor for the oligomeric intermediates.

The test method for measuring mineralization of plastics under composting conditions (ISO 14855) was modified by using activated vermiculite (clay mineral) as a solid matrix instead of mature compost [19]. It was shown that the vermiculite test method is suitable to perform ecotoxicological studies. The advantage of using vermiculite is the possibility of treating it with suitable solvents to obtain clear solutions, suitable for further chemical analyses. The biodegradable plastic material composed of starch, polycaprolactone and polyurethane has been tested in vermiculite and mature compost using the Flash test. It was suggested that the Flash test in combination with the vermiculite biodegradation is a valuable, reliable tool for ecotoxicological assessment of biodegradable plastics. It is worth noting that recently the method based on vermiculite was applied in the EN standard [20].

Ecotoxicity tests of compostable polymers are summarized in Table 7.6.

7.5. CONCLUSIONS

The evaluation of the ecotoxicity of compostable polymers is an important issue to ensure that the produced compost is safe and causes no health hazards.

Several ecotoxicity tests should be used to assess the potential toxic activity against plant and animal life during their degradation and after they are degraded.

Table 7.6. Biodegradable polymers ecotoxicity testing

Polymers	Materials description	Methods	Test species	Test conditions	References
Bionolle™– starch blend	Starch–polyester blend	Modified ASTM E 1676-97	Earthworms *Eisenia foetida*	Compost; 42 days; external temperature: 22°C	12
Starch	J.T. Baker No. 4010	OECD DIN	• Seeds of higher plants (cress, rape, millet) • Earthworms *Eisenia foetida* • *Daphnia magna* • *Vibrio fisheri*	Soil; 160 days; outdoor conditions	13
Cellulose	Fluka No. 22181	OECD DIN	• Seeds of higher plants (cress, rape, millet) • Earthworms *Eisenia foetida* • *Daphnia magna* • *Vibrio fisheri*	Soil; 160 days; outdoor conditions	13
Lignocell BK-40–90	Wood in the form of sawdust; J. Rettenmaier	OECD DIN	• Seeds of higher plants (cress, rape, millet) • Earthworms *Eisenia foetida* • *Daphnia magna* • *Vibrio fisheri*	Soil; 160 days; outdoor conditions	13
FASAL F129	Biogenic material based on the combination of starch with wood; IFA-Tulln	OECD DIN	• Seeds of higher plants (cress, rape, millet) • Earthworms *Eisenia foetida* • *Daphnia magna* • *Vibrio fisheri*	Soil; 160 days; outdoor conditions	13
ÖKOPUR	Biogenic material based on the combination of	OECD DIN	• Seeds of higher plants (cress, rape, millet)	Soil; 160 days; outdoor conditions	13

Abbrev.	Material	Test method	Test organisms	Test conditions	Ref.
	starch with sugar beet residues; IFA-Tulln		• Earthworms *Eisenia foetida* • *Daphnia magna* • *Vibrio fisheri*		
BAK 1095	Synthetic poly (esteramide); Bayer	OECD DIN	• Seeds of higher plants (cress, rape, millet) • Earthworms *Eisenia foetida* • *Daphnia magna* • Luminescent bacteria *Vibrio fisheri*	Soil; 160 days; outdoor conditions	13
PLA	Poly(lactic acid); laboratory synthesized	OECD 208 ISO 11348-3 the Flash test	• Seeds of plants (cress, radish, barley) • Luminescent bacteria *Vibrio fisheri*	Compost, 200 days (controlled composting test; according to EN 14046)	14
PEA	Poly(esteramide); laboratory synthesized	OECD 208 ISO 11348-3 the Flash test	• Seeds of plants (cress, radish, barley) • Luminescent bacteria *Vibrio fisheri*	Compost, 200 days (controlled composting test; according to EN 14046)	14
PEU	Poly(esterurethane); laboratory synthesized	OECD 208 ISO 11348-3 the Flash test	• Seeds of plants (cress, radish, barley) • Luminescent bacteria *Vibrio fisheri*	Compost, 200 days, (controlled composting test; according to EN 14046)	14
PEU	Lactic acid-based poly (esterurethane); laboratory synthesized	OECD 208 the Flash test	• Seeds of plants (cress, radish, barley) • Luminescent bacteria *Vibrio fisheri*	Compost, 112 days, 58°C (controlled composting test)	15
Ecoflex	Aliphatic–aromatic copolyester	DIN 38412 part 30 DIN 38412 part 34	• *Daphnia* • Luminescent bacteria	Synthetic medium + thermophilic strain *T. fusca*, 21 days, 55°C	16
2030/489	Experimental product composed of starch, polycaprolactone, poly-urethane (Estane 54351)	The Flash test	Luminescent bacteria	Composting in activated vermiculite	17

REFERENCES

[1]   Bastioli C.: Properties and applications of Mater-Bi starch-based materials, *Polym. Deg. Stab.* 59 (1998) 263.

[2]   EN 13432: 2000 Packaging – Requirements for packaging recoverable through composting and biodegradation – Test scheme and evaluation criteria for the final acceptance of packaging.

[3]   ISO 14855-1:2006 Determination of the ultimate aerobic biodegradability of plastic materials under controlled composting conditions – Method by analysis of evolved carbon dioxide – Part 1: General method.

[4]   ISO 14851:1999 Determination of the ultimate aerobic biodegradability of plastic materials in an aqueous medium – Method by measuring the oxygen demand in a closed respirometer.

[5]   ISO 14852:1999 Determination of the ultimate aerobic biodegradability of plastic materials in an aqueous medium – Method by analysis of evolved carbon dioxide.

[6]   ASTM D6400–04, Standard specification for compostable plastics, ASTM International.

[7]   Pagga U.: Biodegradability and compostability of polymeric materials in the context of the European packaging regulation, *Polym. Degrad. Stab.* 59 (1998) 371.

[8]   De Wilde B., Boelens J.: Prerequisites for biodegradable plastic materials for acceptance in real-life composting plants and technical aspects, *Polym. Degrad. Stab.* 59 (1998) 7.

[9]   Biodegradable plastics – Developments and environmental impacts, Nolan-ITU Pty Ltd, Environment Australia, October 2002.

[10] Kapanen A., Itävaara M.: Ecotoxicity tests for compost applications, *Ecotoxicol. Environ. Safety* 49 (2001) 1.

[11] Jennings V.L.K., Rayner-Brandes M.H., Bird D.J.: Assessing chemical toxicity with the biolumin-escent photobacterium (*Vibrio fischeri*): a comparison of three commercial systems, *Wat. Res.* 35 (2001) 3448.

[12] Shettlemore M.G., Bundy K.J.: Examination of in vivo influences on bioluminescent microbial assessment of corrosion product toxicity, *Biomaterials* 22 (2001) 2215.

[13] Lappalainen J., Juvonen R., Vaajasaari K., Karp M.: A new Flash method for measuring the toxic-ity of solid and colored samples, *Chemoshpere* 38 (1999) 1069.

[14] Jayeskara R., Sheridan S., Lourbakos E., Beh H., Christie G.B.Y., Jenkins M., Halley P.B., McGlashan S., Lonergan G.T.: Biodegradation and ecotoxicity evaluation of a Bionolle and starch blend and its degradation products in compost, *Ind. Biodeter. & Biodeg.* 51 (2003) 77.

[15] Fritz J., Sandhofer M., Stacher C., Braun R.: Strategies for detecting ecotoxicological effects of biodegradable polymers in agricultural applications, *Macromol. Symp.* 197 (2003) 397.

[16] Tuominen J., Kylmä J., Kapanen A., Venelampi O., Itävaara M., Seppälä J.: Biodegradation of lac-tic acid based polymers under controlled composting conditions and evaluation of the ecotoxico-logical impact, *Biomacromolecules* 3 (2002) 445.

[17] Kapanen A., Venelampi O., Vikman M., Itävaara M.: Testing the ecotoxicity of biodegradable plastics, KTBL-Schrift 414: biodegradable materials and natural fibre composites. Darmstadt: Kuratorium für Technik und Bauwesen in der Landwirtschaft e.V. (KTBL), 2002, 107.

[18] Witt U., Einig T., Yamamoto M., Kleeberg I., Deckwer W.-D., Müller R.-J.: Biodegradation of aliphatic-aromatic copolyesters: evaluation of the final biodegradability and ecotoxicological impact of degradation intermediates, *Chemosphere* 44 (2001) 289.

[19] Degli-Innocenti F., Bellia G., Tosin M., Kapanen A., Itävaara M.: Detection of toxicity released by biodegradable plastics after composting in activated vermiculite, *Polym. Degrad. Stab.* 73 (2001) 101.

[20] ISO 14855:1999/Amd1:2004 Use of activated vermiculite instead of mature compost.

Chapter 8

Environmental impact of compostable polymer materials

Chapter 8
Environmental impact of compostable polymer materials

8.1. INTRODUCTION

Overconsumption, resource utilization, pollution and overpopulation are given as examples of the most basic problems for the environment in the future [1]. A more sustainable future can be achieved by producing more sustainable products causing fewer environmental impacts. A sustainable product is a product that has as little an impact on the environment as possible during its life cycle [1]. The life cycle in this simple definition includes extraction of raw material, production, use and final recycling (or deposition). There are many alternative approaches to look at regarding the environmental impact of plastics, including compostable polymer materials in different applications. Some important methods for evaluating environmental impacts include: life cycle assessment (LCA), Eco-Indicator'99, Environmental Priority System (EPS) and Tellus [1]. The LCA methodology is probably the most widespread technique for evaluating environmental impacts associated with material products.

The environmental impact can be divided to:

- global
- regional
- local

effects.

Used parameters in environmental impact analysis include resource consumption, energy consumption, total waste production, greenhouse gas emissions, regulated air pollutants release, water discharges, etc. [2].

8.2. LIFE CYCLE ASSESSMENT METHODOLOGY

Life cycle assessment (LCA) is a methodological framework for estimating and assessing the environmental impacts attributable to the life cycle of a product, such as climate change, stratospheric ozone depletion, tropospheric ozone (smog) creation, eutrophication, acidification, toxicological stress on human health and ecosystems, the depletion of resources, water use, land use, noise, and others [3, 4].

LCA is a useful tool for measuring environmental sustainability and identifying environmental performance-improvement objectives [5]. Environmental sustainability is about making products that serve useful market and societal functions with less environmental impact than currently available alternatives. Moreover, environmental sustainability necessarily implies a commitment to continuous improvement in environmental performance. The key measurement tool for environmental sustainability is life cycle assessment.

LCA is now a widely acknowledged approach to characterize the environmental impact of products and processes, and its methodology has been standardized under the ISO 14040

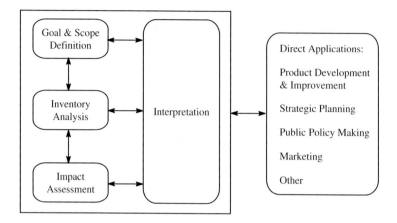

Figure 8-1 Phases of life cycle assessment (LCA).

series [6–9]. Life cycle analysis is used to evaluate environmental impact and potential factors related to product life cycle energy balance, including raw materials, production, consumption, and waste utilization. Life cycle assessment is a method to quantify a product's or system's environmental impact over its life cycle. The LCA method evaluates impact on a global scale. Examples of LCA applications are comparisons of different product and system concepts, or different materials, production or recycling methods. LCA can be used as a tool to detect potentials for improvements with the aim of reducing the impact on human health, the natural environment and resource depletion.

Basically, the LCA method consists of the following steps (Fig. 8.1):

- Goal and scope definition
- Inventory analysis
- Impact assessment
- Interpretation, i.e. reporting and suggestions for product improvement

The goal definition stage defines the purpose, scope, and boundaries of the study, the functional unit, key assumptions to be made and likely limitation of the work [10]. The goal and scope definition of an LCA provides a description of the product system in terms of the system boundaries and a functional unit [3]. The functional unit is the important basis that enables alternative goods, or services, to be compared and analysed. Practitioners may compare, for example, alternative types of packaging on the basis of $1\,\text{m}^3$ of packed and delivered product – the service that the product provides.

Life cycle inventory (LCI) is a methodology for estimating the consumption of resources and the quantities of waste flows and emissions caused or otherwise attributable to a product's life cycle [3]. The inventory analysis constitutes a detailed compilation of all of the environmental inputs and outputs to each stage of the life cycle [10]. The inventory usually includes raw material and energy consumed, emissions to air and water, and solid waste produced. The processes within the life cycle and the associated material and energy flows as well as other exchanges are modelled to represent the product system and its total inputs and outputs from and to the natural environment, respectively (Fig. 8.2). This results in a product system model and an inventory of environmental exchanges related to the functional unit.

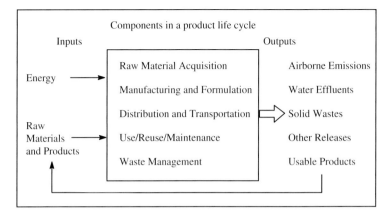

Figure 8-2 Life cycle inventory.

Life cycle impact assessment (LCIA) is a process whereby environmental impacts from the inventory are assessed, and generally the overall environmental performance of the product is determined. LCIA provides indicators and the basis for analysing the potential contributions of the resource extractions and wastes/emissions in an inventory to a number of potential impacts. The result of the LCIA is an evaluation of a product life cycle, on a functional unit basis, in terms of several impact categories (such as climate change, toxicological stress, noise, land use, etc.) and, in some cases, in an aggregated way (such as years of human life lost due to climate change, carcinogenic effects, noise, etc.) [3]. According to standard ISO 14042 life cycle impact assessment consists of two mandatory elements, classification and characterization, and a series of optional elements, normalization, grouping, and weighting [5, 8].

Life cycle interpretation occurs at every stage in an LCA [3]. It serves to evaluate the study in order to derive recommendations and conclusions. The last component of an LCA is to find the ways to improve or to redesign the production processes, or to reduce the costs and the materials used. Several relevant aspects, such as environmental, financial, convenience, and safety, are usually incorporated for improvement assessment or interpretation.

LCA studies that cover cradle-to-grave assessments based on a "functional unit" of product and are conducted in accordance with ISO 14040 series (or equivalent best-practice guidance) are "full" LCAs [10]. Where studies are limited, for example by assessments being of a cradle-to-factory gate nature or where only very limited environmental impact categories are used, e.g. greenhouse gases only, then they are considered to be "partial" LCAs.

Life cycle assessment assists in [2]:

- Identifying opportunities to improve the environmental aspects of products or processes at various points in their life cycle (e.g. strategic planning, priority)
- Decision-making in industry, government or non-governmental organizations for strategic planning purposes, and improving the overall environmental performance and economic performance
- Selection of relevant indicators of environmental performance, including measurement techniques
- Marketing (for example, environmental claims, an ecolabelling scheme or environmental product declaration)

8.3. LIFE CYCLE ASSESSMENT OF POLY(LACTIC ACID)

Literature concerning LCA studies of poly(lactide acid) (PLA) is rather scarce. Vink *et al.* gave an overview of applications of LCA to polylactide acid (PLA) production and provided insight into how they are utilized [5]. Two systems boundaries are depicted: PLA production by Cargill Dow at the Blair, Nebraska, facility and the next-generation PLA facility (five years) with biomass feedstock and wind energy input.

The analysis of PLA production systems (referred to as PLA-Year 1) includes impacts associated with:

- Corn growing
- Transport of corn to the corn wet mill
- Processing of corn to dextrose
- Conversion of dextrose into lactic acid
- Conversion of lactic acid into lactide and
- Polymerization of lactic acid into polylactide

Corn growing includes inputs such as corn seed, fertilizers, electricity and fuel used by farmers, atmospheric carbon dioxide take-up through the photosynthesis process, irrigation water and pesticides. On the output side, emissions such as dinitrogenoxide, nitrates and phosphates were taken into account. Energy and operating supplies (such as process water, cooling water, nitrogen, compressed air, catalysts, stabilizers and chemicals) are all accounted for in the ecoprofile.

The next-generation PLA facility (referred as to PLA-Year 5) differs from PLA-Year 1 in improvements and changes leading to lower fossil fuel and raw material use as well as lower air emissions, water emissions and solid waste production. For example, instead of corn-derived dextrose, the primary feedstock is crop residue (stems, straw, husks and leaves) from corn or other crops or instead of electricity from the Nebraska grid wind power will be used as additional electricity inputs.

For a comparison of PLA ecoprofiles with traditional petrochemical-based polymers the same methodology, software, and core databases were developed as used in the Association of Plastics Manufacturers of Europe (APME) analyses. The APME has over the last ten years published a series of ecoprofiles for traditional petrochemical-based polymers [12].

The study presented by Vink *et al.* was focused on the environmental performance of PLA as measured by three life cycle impact categories: fossil energy requirement, greenhouse gases and water use. The results are presented in Table 8.1.

Table 8.1. Cradle-to-factory gate energy use and CO_2 of PLA [5, 11]

	Process energy, fossil, GJ/t plastic	Feedstock energy, fossil, GJ/t plastic	Total fossil energy, GJ/t plastic	Fossil CO_2 from process energy, kg/t plastic	CO_2 absorption, plant growth, kg/t plastic	Net CO_2, kg/t PLA*
PLA-Year 1	54	0	54	3450	−2190	1260
PLA- Target. Year 5	7	0	7	520	−2280	−1760
HDPE	31	49	80	1700	0	1700
PET (bottle grade)	38	39	77	4300	0	4300
Nylon 6	81	39	120	5500	0	5500

* Equals the sum of "Fossil CO_2 from process energy" (positive value) and "CO_2 absorption, plant growth" (kg/t plastic).

The gross fossil energy requirement is 54.1 MJ/kg of PLA. The polylactide production system (PLA-Year 1) uses 22–55% less fossil energy than the petroleum-based polymers. With the process improvements targeted by Cargill Dow (PLA-Year 5) the use of fossil energy can be reduced by more than 90% compared to any of the petroleum-based polymers being replaced [5].

The contribution to global climate change, identified as perhaps the most important environmental issue of this century, was performed on the basis of greenhouse gas emissions. As in the comparison of fossil energy use, the analysis compared conventional polymers with PLA from cradle to pellet (from raw materials to the point where the product is ready for shipment to a converter or fabricator). All emissions were converted to CO_2 equivalents in order to facilitate the comparison. Life cycle assessment revealed that the PLA-Year 1 production process enjoyed a substantial advantage over most polymers, and was comparable to several others. A targeted PLA process (PLA-Year 5) resulted in significant greenhouse gas emissions performance improvements, i.e. -1.7 kg-CO_2 eq./kg PLA.

Regarding water use, it was reported that the total amount of water required was competitive with the best performing petrochemical polymers.

The estimated cradle-to-factory gate energy requirements for PLA production from rye and whey in Table 8.2 show that also small plants (3 kt p.a. and 4.2 kt p.a., respectively) may be rather energy efficient (the estimated values remain to be proven in commercial plants) [11, 13].

Table 8.2. Cradle-to-factory gate energy use of PLA [11, 13]

	Fossil energy in GJ/t PLA		
	from rye*	from whey**	Cargill Dow***
Cultivation	8.7	0.0	
Milling	6.6	0.0	
Transportation	0.0	2.3	
Hydrolysis and fermentation	33.9	25.0	
Polymerization	12.8	12.8	
Cradle-to factory gate energy	**62.1**	**40.1**	**54**

* Based on a 3 kt PLA plant according to Inventa-Fischer (preliminary).
** Based on a 4.2 kt PLA plant according to Fraunhofer-IGB.
*** Cargill Dow (Year 1).

8.4. POLYHYDROXYALKANOATES

Contrary to the environmental analyses for PLA and starch polymers, the results for polyhydroxyalkanoates (PHA) are based on simulations since no large-scale facility is available to date [13]. Various studies reported for environmental impact of PHAs differ widely. Gerngross and Slater suggested that PHAs may not necessarily be environmentally friendly in view of fossil fuel consumption [14–16]. In terms of land use, resource depletion and emission to air and water PHA production by fermentation score worse than conventional polymer production [16]. For example, it was reported that the amount of fossil fuel (2.39 kg) required to produce 1 kg of PHAs exceeded that required (2.26 kg) to produce an equal amount of polystyrene [14]. The global warming potential associated with the life cycle of polyhydroxyalkanoate produced in genetically engineered corn developed by Monsanto was assessed by Kurdikar *et al.* [17]. In this study, the grain corn is harvested in a conventional manner, and the polymer is extracted from the corn stover (i.e. residues such as stalks, leaves, and cobs), which would be otherwise left on the field. While corn farming was assessed based on current practice, four different hypothetical PHA production scenarios were tested for the extraction process. Each scenario differed in the energy source used for polymer extraction and compounding, and the results were compared to polyethylene (PE). The first scenario involved burning of the residual biomass (primarily cellulose) remaining after the polymer was extracted from the stover. In three other scenarios, the use of conventional energy sources of coal, oil and natural gas was investigated. It was indicated that an integrated system, wherein biomass energy from corn stover provides energy for polymer processing, would result in a better greenhouse profile for PHA than for PE. However, plant-based PHA production using fossil energy sources provides no greenhouse gas advantage over PE, in fact scoring worse than PE. It is noteworthy that the results are based on a cradle-to-pellet modelling, and not cradle-to-grave.

Table 8.3. Cradle-to-factory gate energy use of PHA [11]

	Total fossil energy (cradle-to-factory gate), GJ/t plastic	Source
PHA grown in corn plants	90	[15]
PHA by bacterial fermentation	81	[15]
PHA grown in corn plants	(100)	[17]
PHA by bacterial fermentation	66–573	[18]
PHA, process unknown	45–65	MITI (2001)
HDPE	80	APME (1999)
PET (bottle grade)	77	APME (1999)
PS (general purpose)	87	APME (1999)

Large-scale fermentative production (amounting to 5000 tonnes per year) of poly(3-hydroxybutyrate-*co*-5 mol% 3-hydroxyhexanoate) (P(3 HB-*co*-5 mol% 3 HHx) from soybean oil as sole carbon source was simulated using a recombinant strain of *Ralstonia eutropha* harbouring a polyhydroxyalkanoate (PHA) synthase gene from *Aeromonas caviae* [19]. Life cycle inventories of energy consumption and carbon dioxide emissions from the cradle to the fermentation factory gate were calculated for the P(3 HB-*co*-5 mol% 3 HHx) copolyester production to examine the basic environmental impact of the production. The corresponding life cycle inventories for fermentative production of poly(3-hydroxybutyrate) (P(3 HB)) was also estimated and used for comparison. In addition, the life cycle inventories of bio-based PHA polymers were compared with those of typical petrochemical-based polymers based on data published by the Association of Plastic Manufacturers in Europe (APME).

The LCI values of energy consumption and CO_2 emissions were estimated to be smaller for the PHA copolymer from soybean oil than for P(3 HB) from glucose (Table 8.4). It was concluded that the life cycle inventories of energy consumption and carbon dioxide emissions of

PHA polymers were markedly lower than those of typical petrochemical polymers, i.e. LDPE, HDPE, PP, PS and PET (bottle grade). Feedstock energy accounted for about 50% or more of cumulative energy use for petrochemicals but nothing for PHAs due to different origin of feedstock and combustible fossils vs renewable plant resources. On the contrary, process energy shares nearly all of the cumulative energy used for PHA production. As for cumulative CO_2 emissions, absorption of CO_2 from the air by the soybean or corn plant has mainly contributed to reducing the CO_2 emissions for PHA production.

Table 8.4. Estimation of total energy and CO_2 emissions for PHAs (from the cradle-to-granules) and for petrochemical polymers (from the cradle-to-pellets) [19]

	Total energy, MJ/kg plastic	CO_2 kg/kg plastic
P(3 HA)	50	0.26
P(3 HB)	59	0.45
LDPE (low density polyethylene)	81	1.9
HDPE (high density polyethylene)	80	1.7
PP (isotactic polypropylene)	77	1.9
PS (polystyrene)	87	2.6
PET (bottle grade) poly(ethylene terephthalate)	79	3.1

Kim and Dale [20] presented a study to estimate the environmental performance of polyhydroxyalkanoates (PHAs), from agricultural production through the PHA fermentation and recovery process – "cradle to gate". Two types of PHA production systems were investigated: corn grain-based PHA (the reference system) and corn grain and corn stover-based PHA (called an "integrated system"). The environmental performance of the PHA production system was compared to that of a conventional polymer fulfilling an equivalent function (i.e. packaging film). The function of the product system was defined as polymer used in a packaging film, 1 kg of PHA resin chosen as the reference flow. The system boundary in the reference system included corn production, dextrose production (corn wet milling), PHA fermentation and recovery process, and up-stream processes (e.g. fertilizers, agrochemicals, fuels, electricity, etc.). In addition to the system boundary of the corn grain-based PHA system, processes for producing PHA derived from corn stover, a crop residue (e.g. harvesting and transporting corn stover, PHA production process from corn stover) were included in the system boundary in the integrated system. Most processes were assumed to occur in the United States. The environmental performance was addressed as non-renewable energy and selected potential environmental impacts including global warming, photochemical smog, acidification, and eutrophication.

Non-renewable energy ranged from 69 to 107 MJ/kg, depending on the PHA fermentation technologies. Global warming associated with corn grain-based PHA was 1.6–4.1 kg-CO_2 eq./kg, demonstrating that changing the PHA fermentation technologies reduces GHG emissions by up to 62%. The lowest global warming occurred in the PHA fermentation technology given by Akiyama *et al.* [19], which also featured the lowest electricity consumption. The PHA fermentation technology given by Gerngross [14] had the highest global warming due to low yield of PHA and high electricity consumption.

The primary contributing process to most environmental impacts except for photochemical smog and eutrophication was the PHA fermentation and recovery process. For photochemical smog and eutrophication, the primary contributing process was corn cultivation due to nitrogen related burdens from soil. It was concluded that the trend of PHA fermentation development showed that the PHA fermentation technology was still immature and continued to improve, thereby also decreasing the environmental impacts.

Even though utilizing corn stover requires more non-renewable energy in harvesting and transporting corn stover, the integrated system (corn grain and corn stover utilized as a raw material for PHA production) can conserve non-renewable energy by 70–84% when compared to the reference system (corn grain-based PHA production system). PHA produced in an integrated system, in which corn stover was harvested and used as raw material for PHA along with corn grain, offered global warming credits (negative greenhouse gas emissions), ranging from -0.28 to -1.9 kg-CO_2 eq./kg, depending on the PHA fermentation technologies employed and significantly reduced the environmental impacts compared to corn-based PHA. The significant reductions from the integrated systems are due to: (1) fewer environmental impacts in corn cultivation and wet milling, and (2) exporting surplus energy from lignin-rich residues in the corn stover process. The integrated system can reduce photochemical smog, acidification and eutrophication, compared to the reference system (corn grain-based PHA production system).

Table 8.5. Non-renewable energy in the corn grain-based PHA (reference system) and an integrated system [20]

Reference system Overall non-renewable energy, MJ/kg	Integrated system Overall non-renewable energy, MJ/kg	Source
107	31.5	Gerngross
	24.9	Metabolix (current)
	12.3	Akiyama
	17.8	Metabolix (near future)

Table 8.6. Global warming associated with corn grain-based PHA (reference system) and an integrated system [20]

Reference system Global warming, kg-CO_2 eq./kg	Integrated system Global warming, kg-CO_2 eq./kg	Source
4.1	-0.28	Gerngross
	-0.77	Metabolix (current)
1.6	-1.93	Akiyama
	-1.19	Metabolix (near future)

Table 8.7. Other environmental impacts with corn grain-based PHA (reference system) and an integrated system

Environmental impact	Reference system	Integrated system	Source
Photochemical smog,	30.7	16.4	Gerngross
mg-NO_x eq. m^{-1}kg^{-1}	27.7	14.6	Metabolix (current)
	22.5	10.2	Akiyama
	20.6	11.7	Metabolix (near future)
Acidification, moles	2.41	0.97	Gerngross
H$^+$ eq. kg^{-1}	2.14	0.81	Metabolix (current)
	1.62	0.36	Akiyama
	1.56	0.62	Metabolix (near future)
Eutrophication,	2.02	1.21	Gerngross
g-N eq. kg^{-1}	1.90	1.14	Metabolix (current)
	1.68	0.98	Akiyama
	1.43	0.94	Metabolix (near future)

Kim and Dale concluded that under the current PHA fermentation technology, corn grain-based PHA did not provide an environmental advantage over polystyrene. However, it was stressed that corn grain-based PHA produced by the near future PHA fermentation technology would be more favourable than polystyrene in terms of non-renewable energy and global warming due to improvement in the PHA fermentation and recovery process. In their opinion, corn grain-based PHA produced even in the near future technology does not provide better profiles for other environmental impacts (i.e. photochemical smog, acidification and eutrophication) than polystyrene. One of the primary reasons for high impacts of PHA in photochemical smog, acidification and eutrophication is the environmental burdens associated with corn cultivation. On the other hand, the integrated system could produce PHA that provides much smaller environmental impacts (except eutrophication) than polystyrene.

8.5. STARCH-BASED POLYMERS

The first publicly available LCA report for bio-based was prepared by Dinkel *et al.* [21] for the Swiss Federal Agency for the Environment and concerned starch-based polymers [22]. The system studied covered the entire production process (cradle-to-factory gate) and the waste management stage. It was concluded that thermoplastic starch performs better than low density polyethylene (LDPE) in all categories, i.e. energy resources, greenhouse gas (GHG) emissions, human toxicity and salinization, except for eutrophication. Results of environmental assessments for starch polymer pellets with different shares of petrochemical polymers are summarized in Tables 8.8 and 8.9 [11, 13, 22]. It was assumed that both the starch polymers and polyethylene are incinerated in municipal solid waste incineration plants after their useful life. No credits have been assigned to steam and/or electricity generated in waste-to-energy facilities. The environmental impact of starch polymers generally decreases with lower shares of petrochemical copolymers [13]. Starch polymers offer saving potentials relative to polyethylene in the range of 24–52 GJ/t plastic and 1.2–3.7 t CO_2/t plastic depending on the share of petrochemical copolymers.

Table 8.8. Cradle-to-factory gate energy use of starch-based polymers [11]

	Total fossil energy (cradle-to-factory gate), GJ/t plastic	Source
TPS	25	[21]
TPS + 15% PVOH	27	[23]
TPS + 52.5% PCL	52	[23]
TPS + 60% PCL	56	[23]
HDPE	80	[11]
50% LLDPE + 50% HDPE	76	[13]

Table 8.9. Greenhouse gas emission from starch-based polymers [13]

	GHG emission, kg-CO_2 eq./kg[*]	Source
TPS	1.1	[21]
TPS + 15% PVOH	1.7	[23]
TPS + 52.5% PCL	3.3	[23]
TPS + 60% PCL	3.6	[23]
50% LLDPE + 50% HDPE	4.8	[13]

[*] Emission refers to incineration in all cases. Exception: composting has been assumed for thermoplastic starch (TPS).

8.6. BLENDS

Novamont applied LCA methodology to evaluate the overall environmental impact due to production and disposal of Mater-Bi bags used by households to collect organic waste in Switzerland [24, 25]. The results of this first analysis of the environmental impact of a Mater-Bi bag performed in comparison with a polyethylene and a paper bag showed that production of paper bags, due to their higher weight, consumed much more energy than production of Mater-Bi and polyethylene bags [24]. On the other hand, the Mater-Bi bag gave an important contribution to the reduction of the greenhouse effect, the effect being evaluated as four times lower than PE bags and five times lower than paper bags, thanks to the presence of its natural components.

The life cycle included raw material acquisition, the production and processing and/or disposal of bags as well as routes of transport. Packaging, distribution, utilization and collection as well as transport to wholesalers could not be considered due to dependency of these processes on the prospective bulk buyers and retailers [25]. Paper bags "haushalt compost", which can be composted, and PE multipurpose bags, which cannot be composted, were used as points of reference. It is not relevant to the overall results whether maize produced in Switzerland or in France is used. Mater-Bi bags made of French maize were selected for the overall assessment as maize on the European market is mainly produced in France.

Environmental impact categories considered in the analysis [26]:

- Energy: consumption of energy resources (oil, natural gas, etc.), assessed from the energy content of the resources necessary (MJ)
- Greenhouse effect: temperature increase of the planet due to gas emissions (CO_2 equivalents)

Table 8.10. Comparison of Mater-Bi[TM] bags with paper bags and bags made of PE [26]

Environmental impact category	Bag made of Mater-Bi[TM] compared with:		
	paper bag	bag made of PE	bag made of PE, including incineration of the organic residue
Energy	+ +	0	+
Greenhouse effect	+	+	+ +
Acidification	+	0	+ +
Nutrification	+ +	0	+
Ozone formation	+ +	+	+ +
Toxicity in air	+	+ +	+ +
Toxicity in water	+ +	0	+
Saltification	−	− −	+ +
Waste produced	+ +	− −	−

Legend:
+ + = much better
+ = better
0 = comparable
− = worse
− − = much worse

- Acidification: potential damage to plants due to the emission of substances such as nitrogen and sulphur oxides (SO_2 equivalents)
- Nutrification: potential unbalancing water and soil due to the emission of substances that have a fertilizing effect, such as nitrates and ammonia (PO_4 equivalents)
- Ozone formation: increase in the formation of ozone (summer smog) due to the emission of substances such as organic solvents and nitrogen oxides (C_2H_4 equivalents)
- Toxicity in air: pollution of the atmosphere due to gas emissions
- Toxicity in water: pollution of water due to organic emissions, heavy metals, etc.
- Salification: damage to flora and fauna in water due to the emission of salts, such as chlorides (assessed as H^+ ions)
- Waste produced: quantity of waste disposed of, weighed as inert substances, harmful toxicity waste, radioactive waste, etc.

The results are summarized in Table 8.10.

The results indicated that the Mater-Bi bag and the multipurpose PE bag can be regarded as equivalent, as long as the focus remains on production and disposal (disregarding compostable waste incineration) [25]. If the compostable waste that is incinerated with PE bags is taken into account, the Mater-Bi bag offers better ecological value. The production and disposal of the paper bag is bound to cause considerably more damage to the environment than that of the Mater-Bi bag. It was concluded that for the municipal collection of organic waste biodegradable bags should be recommended.

Life cycle assessments were also applied to analyse the environmental effects related to the production and disposal of loose fill made out of Mater-Bi pellets in comparison to those made by expanded polystyrene (EPS) [25]. The life cycle included raw material acquisition, the production and processing and/or disposal of bags as well as routes of transport. Packaging, distribution, utilization and collection as well as transport to wholesalers could not be considered due to dependency of these processes on the prospective bulk buyers and retailers. Life cycle profiles were drawn up using the modified-impact-oriented model and the impact categories of Eco-Indicator'95. Thirteen different impact categories were taken to analysis, including energy, global warming, and acidification. In eight of these 13 impact categories the production and disposal of Mater-Bi loose fill causes less environmental damage than EPS loose fill [22, 25]. The environmental impact of Mater-Bi loose fill is reported to be significantly lower for the categories of winter smog, air toxicity and carcinogeneity. The impact of Mater-Bi loose fills is lower than EPS loose fills with regard to energy use, global warming, acidification, ozone creation/summer smog and heavy metals. In two categories, Mater-Bi loose fill has a larger environmental impact than EPS loose fill (salinization and deposited waste) while the effects are comparable for the three remaining categories (eutrophication, toxicity water and ozone layer depletion). The general conclusion was that from the ecological point of view the Mater-Bi loose fills have to be given preference over EPS loose fills.

A streamlined life cycle assessment (LCA) was undertaken on a selection of degradable plastics application for film blowing into shopping bags in Australia [27]. The degradable plastic materials that are suitable for applications in film blowing for shopping bags and/or currently available on the market were studied, including the following compostable polymer materials: blends of maize starch with polycaprolactone (e.g. Mater-Bi), blends of maize starch (50%) and aliphatic polyesters (e.g. Ecoflex, Bionolle) and polylactic acid. For the purpose of the study, the "functional unit" has been defined as a household carrying approximately 70 grocery items home from a supermarket each week for 52 weeks. Several different waste management technologies were modelled, i.e. landfill (anaerobic environment), source separated green and food Mechanical Biological Treatment (MBT) composting, municipal solid waste (MSW) composting and municipal solid waste anaerobic digestion. Assumptions were made to model the baseline end-of-life waste management destinations of the degradable plastics, i.e. 70.5% of degradable bags go to landfill, 10% of degradable bags go to composting (source separated organics), 19% of degradable bags are reused and 0.5% of degradable bags end up as litter. The environmental impact categories analysed in the impact assessment include: greenhouse, resource depletion, eutrophication, litter aesthetics and litter biodiversity. The results are compared with conventional HDPE, paper bags, reusable plastic bags and calico bags. The conclusion was that reusable bags have lower environmental impacts than all of the single-use bags, including both conventional HDPE and degradable bags.

8.7. OVERVIEW

An overview of the life cycle inventory for compostable polymer materials is given in Tables 8.11, 8.12 and 8.13. The cradle-to-factory gate energy requirements for PLA are 20–30% below those for polyethylene, while greenhouse gas (GHG) emissions are about 15%–25% lower. For starch polymer pellets energy requirements are mostly 25–75% below those for polyethylene (PE) and greenhouse gas (GHG) emissions are 20–80% lower. Except for eutrophication starch polymers (both TPS and copolymers) score better than PE also for all other indicators covered by the LCA [13]. The environmental impact of starch polymers generally decreases with lower shares of petrochemical copolymers. The results for PHA vary greatly. LCA results are only available for energy use.

Table 8.11. Key life cycle inventory results for some compostable polymer materials [1, 22]

	Compostable polymers based on renewable feedstock				Compostable polymers based on petrochemical feedstock		Petrochemical polymers		
	PLA	Starch-based (Mater-Bi)	TPS	PHA	PCL	PVA	HDPE	LDPE	LDPE
Cradle-to-gate non-renewable energy, GJ/t	54.2	53.5	25.4	81* 66–573**	83 77	102 58	79.9	80.6	91.7
Emission of GHG, kg-CO$_2$ eq./kg	3.45	1.21	1.14	n/a	3.1*** 5.0–5.7***	2.7*** 4.1–4.3***	4.84	5.04***	5.2***
Type of waste treatment assumed for calculation of emissions	Incineration	Composting	Incineration; composting	n/a	Incineration	Incineration	Incineration	Incineration	80% incineration +20% landfilling

* PHA by fermentation; ** PHA, various processes; *** only CO$_2$. Embodied carbon: 3.14 kg CO$_2$/kg PE, 2.32 kg CO$_2$/t PCL, 2.00 kg CO$_2$/t PVA.

Table 8.12. Results for energy and GHG emissions (in %) [11]

	Savings compared to conventional polymers	
	Fossil energy, %	GHG emission, %
1. Pellets (granules)		
TPS	−65 to −75	−75 to −80
TPS + 15% PVOH	−65 to −75	−60 to −70
TPS + 52.5% PCL	−35 to −50	−25 to −35
TPS + 60% PCL	−30 to −40	−20 to −30
PLA	−20 to −30	−15 to −25
PHA	−10 to −20 up to +700	n/a
2. End products, bioplastics		
Starch loose fills	−30 to −40	−60 to +58
Starch films and bags	−50	−60

Table 8.13. Savings relative to petrochemical counterparts [11]

	Energy savings, MJ/kg bio-based polymer*	GHG savings, kg-CO_2 eq./kg bio-based polymer*
TPS	51	3.7
TPS + 15% PVOH	52	3.1
TPS + 52.5% PCL	28	1.4
TPS + 60% PCL	24	1.2
Mater-Bi foam grade	42	3.6
Mater-Bi film grade	23	3.6
PLA	19	1.0

* Max. ±15% depending on whether LDPE or LLDPE, according to APME, is chosen as reference.

LCA data for polycaprolactone (PCL) and polyvinyl alcohol (PVA) are generally considered to be subject to major uncertainties [22]. The life cycle analysis concerning biodegradable polymers, including aliphatic polyester based on petrochemical feedstock, i.e. poly(butylene succinate), was done in Japan under the sponsorship of MITI (Ministry of Trade and Industry) [28]. It was found that poly(butylene succinate) is comparable to the non-biodegradable polyesters in CO_2 emissions, but have clear advantages in the resource and solid waste impact categories.

The environmental effects of substituting bio-based polymers for petrochemical polymers on a large scale were estimated [13]. Two perspectives were taken. First, the savings of fossil fuels, the effects of greenhouse emissions and the consequences for land use (in Europe) were studied. Second, it was analysed whether the lower specific impact of bio-based polymers (e.g. kg-CO_2 eq. per kg of polymer) can (over)compensate the additional environmental impacts caused by expected high growth in petrochemical plastics.

In the study, one mass unit of polymer in primary form was chosen as the basis of comparison. The environmental analyses were conducted based on two types of system boundaries:

- The cradle-to-factory gate
- The cradle-to-grave

The first approach covers the environmental impacts of a system that includes all processes from the extraction of the resources to the product under consideration, i.e. one mass unit of polymer. The second one additionally includes the use phase and the waste management stage. In the study the use phase (including further processing to an end product and its use) was excluded, assumed to be comparable for the various types of polymers studied. The chosen impact categories include energy use, greenhouse gas (GHG) emission and land use. For energy data, cradle-to-factory gate values were used, whereas for GHG emission data, cradle-to-grave data were used.

It was estimated that the energy and emission savings resulting from bio-based polymers were rather high as the comparison with the energy use of other bulk material showed. The lower end of energy savings related to bio-based polymers (calculated on the basis of petrochemical polymers), amounting to 10–15 GJ/t, and was in a similar range as the total energy needed to make 2–3 tonnes of cement, 1–2 tonnes of secondary steel (electric arc steel) or of recycled glass, about 1 tonne of paper/board or ca. 0.5 tonnes of recycled aluminium.

In specific terms, related to mass unit of polymer, bio-based polymers were very attractive in terms of specific energy and emissions savings. Energy and emissions savings in specific terms were found to be 20–50 GJ/t polymer and 1.0–4.0 t-CO_2 eq./t polymer, respectively. It is noteworthy that the data used to estimate the savings were valid for a "cradle-to-grave" system where the waste management technology was incineration without energy recovery. The results of the calculations on land use requirements showed that by 2010 a maximum of 125 000 ha may be used for bio-based polymers in Europe and by 2020 an absolute maximum of 975 000 ha. Comparing this with total land use in Europe (15 countries) for various purposes showed that if all bio-based polymers were to be produced from wheat, land requirements range would be from 1 to 5%, depending on growth scenario.

8.8. CONCLUSIONS

Life cycle assessment (LCA) studies are of increasing importance for compostable polymer materials, regarding the optimization of the process, external communication and promotion, legislative and policy issues. However, the availability of life cycle assessment studies on compostable polymer materials (including bio-based polymers) is still quite limited. Most studies concern only one group of compostable polymer materials, i.e. bio-based polymers. The literature concerning compostable polymers based on petrochemical feedstock is scarce. Moreover, many of the environmental analyses choose the cradle-to-gate perspective (i.e. the analysis ends with the product under consideration) [22]. Additional analyses, taking a cradle-to-grave perspective by including all major waste management options (landfilling, composting, MSWI plants, waste-to-energy facilities, digestion and recycling), are recommended [22].

The most important uncertainties in published LCA studies relate to the waste management phase, especially regarding methane emissions from landfills, energy recovery yields in waste-to-energy facilities and carbon sequestration due to composting [28].

Further full-sized LCA studies for compostable polymer materials are necessary to derive the final conclusions about environmental benefits. Substantial scope for improvement can be expected for optimization of the production process (e.g. PLA) by increasing the efficiencies of the various unit processes involved and by process integration [22]. It is noteworthy that compostable polymer materials are still at the development stage, whereas the manufacture of conventional petrochemical polymers has been optimized for decades.

The available LCA results show that compostable polymers have advantages over petro-chemical-based polymers in several environmental impact categories (including typically fossil energy consumption, greenhouse gas emissions) but are less favourable or poorer in other categories (typically eutrophication, ozone layer depletion, and in some cases acidification). In addition, the different impact categories are usually not regarded as being of equal weighting in terms of seriousness of effect on the environment (e.g. global warming is generally regarded as far more important than eutrophication).

In general, compostable polymer materials are reported to have favourable eco-profiles for many applications due to their relatively low energy in manufacture, their CO_2 "neutral" status, and their end-of-life "value" in composting or energy recovery.

REFERENCES

[1]   Ljunberg L.Y.: Materials selection and design for development of sustainable products, *Materials and Design* 28 (2007) 466.
[2]   Narayan R.: Environmental footprint/profile of biobased, biodegradable products in Workshop: Assessing the sustainability of biobased products, June 26–27, 2003, Iowa State University, USA. Available: ww3.abe.iastate.edu/biobased/LCAfootprint.pdf.
[3]   Rebitzer G., Ekvall T., Frischknecht R., Hunkeler D., Norris G., Rydberg T., Schmidt W.-P., Suh S., Weidema B.P., Pennington D.W.: Life cycle assessment. Part I: Framework, goal and scope definition, inventory analysis, and applications, *Environment International* 30 (2004) 701.
[4]   Pennington D.W., Potting J., Finnveden G., Lindeijer E., Jolliet O., Rydberg T., Rebitzer G.: Life cycle assessment. Part II: Current impact assessment practice, *Environment International* 30 (2004) 721.
[5]   Vink E.T.H., Rábago K.R., Glassner D.A., Gruber P.R.: Applications of life cycle assessment to NatureWorks™ polylactide (PLA) production, *Polym. Deg. Stab.* 80 (2003) 403.
[6]   ISO 14040:1997 – Environmental management – Life cycle assessment – principles and framework.
[7]   ISO 14041:1998 – Environmental management – Life cycle assessment – goal and scope definition and inventory analysis.
[8]   ISO 14042:2000 – Environmental management – Life cycle assessment – life cycle impact assessment.
[9]   ISO 14043:2000 – Environmental management – Life cycle assessment – life cycle interpretation.
[10] Murphy R., Bartle I.: Biodegradable polymers and sustainability: insight from life cycle assessment, Summary Report.
[11] Patel M.: Life cycle assessment of synthetic and biological polyesters, International Symposium on Biological Polyesters, 22–26 September, Münster, Germany.
[12] APME, Association of Plastics Manufacturers in Europe, Avenue E. van Nieuwenhuyse 4, B-1160 Brussels, Belgium, www.apme.org
[13] Techno-economic feasibility of large-scale production of bio-based polymers in Europe (PRO-BIP), Final Report, Utrecht/Karlsruhe, October 2004.
[14] Gerngross T.U.: Can biotechnology move us toward a sustainable society? *Nat. Biotechnol. 1* (1999) 541.
[15] Gerngross T.U., Slater S.: How green are green plastics? *Scientific American*, August 2000, 37.
[16] Gerngross T.U.: Polyhydroxyalkanoates: the answer to sustainable polymer production, *ACS Symposium Series*, 784 (2001) 10.
[17] Kurdikar D., Fournet L., Slater S.C., Paster M., Gerngross T.U., Coulon R.: Greenhouse gas profile of a plastic material derived from a genetically modified plant, *J. Ind. Ecol.* 4 (2001) 107.

[18] Heyde M.: Ecological considerations on the use and production of biosynthetic and synthetic bio-degradable polymers, *Polym. Deg. Stab.* 59 (1998) 3.

[19] Akiyama M., Tsuge T., Doi Y.: Environmental life cycle comparison of polyhydroxyalkanoates produced from renewable carbon resources by bacterial fermentation. *Polym. Deg. Stab.* 80 (2003) 183.

[20] Kim S., Dale B.E.: Life cycle assessment study of biopolymers (polyhydroxyalkanoates) derived from no-tilled corn, *The International Journal of Life Cycle Assessment* 10 (2005) 200.

[21] Dinkel F., Pohl C., Ros M., Waldeck B.: Ökobilanz stärkehatliger Kunstoffe (Nr. 271), 1996, 2 volumes. Study prepared by CARBOTECH, Basel, for the Bundesamt für Umwelt und Landschaft (BUWAL), Bern, Switzerland.

[22] Patel M., Bastioli C., Marini L., Würdinger E.: Environmental assessment of bio-based polymers and natural fibres, in: *Biopolymers*, vol.10 (2003), 409, Wiley-VCH.

[23] Patel M., Jochem E., Marscheider-Weidemann F., Radgen P., von Thienen N.: C-STREAMS – Estimation of material, energy and CO_2 flows for model systems in the context of non-energy use, from a life cycle perspective. Volume 1 (1999). Report by Fraunhofer ISI Karlsruhe, Germany.

[24] Bastioli C.: Global status of the production of biobased packaging materials, *Starch* 53 (2001) 351.

[25] Schwarzwälder B., Estermann R., Marini L.: The part of life-cycle-assessment for biodegradable products: bags and loose life, www.composto.ch/publikationen

[26] www.novamont.com

[27] The impact of degradable plastic bags in Australia. Final Report to Department of the Environment and Heritage, prepared by ExcelPlas Australia, Centre for Design at RMIT, Nolan-ITU, September 2003.

[28] Narayan R., Patel M.: Review and analysis of bio-based product LCAs in Workshop: Assessing the sustainability of biobased products, June 26–27, 2003, Iowa State University, USA. Available: ww3.abe.iastate.edu/biobased/LCAfootprint.pdf.

Chapter 9
Perspectives

Chapter 9
Perspectives

Recently, the main obstacles for developing compostable polymer materials have been reported as price, performance, manufacturing process (scale-up) and legislative issues. Below are some developments in these areas.

9.1. PRICE EVOLUTION

The price evolution of compostable polymer materials over the past 10 years are presented in Table 9.1.

In 1998 the prices of compostable polymers were high. Therefore, it was often stated that

Table 9.1. Evolution of price of compostable polymer materials [1-5]

Polymer	Manufacturer	1998 price, USD/kg	2002 price	Expected
PHB		15	20 EUR/kg	2.4 EUR/kg 2–3 EUR/kg 3.8–4 USD/kg
PHA			10–20 EUR/kg	1–2 EUR/kg 2.5–3 EUR/kg 3.5–4.5 USD/kg
PLA		>20	2.2–3.4 EUR/kg (2.2 EUR/kg, 2004)	1.5–1.8 EUR/kg
PCL		6		
Bionolle (aliphatic polyesters based on succinic acid)		8–10	3.5	
BAK 1095 (aliphatic polyesteramides)		5		
Ecoflex (aliphatic-aromatic polyesters)		5	3.1 EUR(2004)	
Mater-Bi (blends of starch		3–7	2.5–3 EUR/kg	
TPS		3	1 EUR/kg 0.2–0.5 EUR/kg	
Petrochemical polymers				
PET			1–1.5 EUR/kg (2004)	
PE		<1 USD	0.8 EUR/kg (2004)	

"Compared to the prize of commodity plastics as LDPE and PS (prices ca. 1 USD/kg) the differences are too large. Reducing their cost prizes will therefore be the major challenge in the further development of these materials. Estimates have forecasted that up scaling and new production techniques can lower prices to about 7–8 USD/kg for PHB and to about 12–17 USD/kg for PLA before the year 2000. According to Cargill, the manufacturer of PLA, the prize can be brought to 5 USD/kg [1]."

In the biodegradable plastics market, aliphatic polyesters (in the main PLA, the present main competitor of starch thermoplastics) sell for £1.5–2.5 per kilo. Starch itself is very cheap (£0.33 per kilo), but starch-based biopolymers such as Mater-Bi (from Novamont) are more expensive (£2.40–3.4 per kilo) [6].

Annual production of 5000 tonnes of P(3HB-*co*-5 mol% 3HHX) was estimated to cost from 3.5 to 4.5 USD/kg, depending on presumed production performances [4]. Similar scale production of P(3 HB) from glucose was estimated to cost 3.8–4.2 USD/kg.

If it is possible to produce polyhydroxyalkanoates at a cost of 1–2 EUR per kg, a diverse range of potential applications become commercially very attractive. According to some industry specialists, this price range is already feasible with current, large-scale and fully integrated bioreactors and downstream processing technology [5].

A price decline of biodegradable plastics is expected due to wide use and increase of the amount of production from 3.6 EUR/kg at 10000 tonne scale production to 1.1 EUR/kg at 15000000 tonnes [7].

In general, the price difference between compostable polymers and standard plastic has decreased in last decade. It concerns, especially polymers based on renewable resources, e.g. polylactic acid.

9.2. CAPACITY

Manufacturing capacities for compostable polymers increase systematically. In 1990 global capacity for compostable polymer materials was 450 tonnes, in 2000 it increased to 44000 and in 2003 reached above 259000 tonnes [8]. An estimation of existing and planned global capacity of compostable polymer materials production is given in Table 9.2. It is noteworthy that many companies declare the increase in the production capacity, e.g. Cargill Dow.

The world market for biodegradable materials is developing very dynamically. With this significant increase in capacity, the chemical companies, for example BASF, would like to participate in this market growth. Due to the recently passed amendment to the German packaging ordinance, BASF expects activity in the domestic market to pick up. Strong growth in interest in compostable polymers, especially based on renewable resources in general, is observed. Progress in material development takes place, e.g. expanding the range of bioplastics (e.g. Ecovio (BASF)), or new manufacturing capacities (e.g. 20000 tonnes biorefinery, Novamont).

9.3. LEGISLATION INITIATIVES

Recently, different legislative measures have been taken in various countries in the world, in order to support sustainable development.

Table 9.2. Estimation of existing and planned capacity (main leaders)

Compostable polymer materials	Company	Capacity, tonne/year 2003	Capacity planned, tonne/year
Based on renewable raw materials			
PLA			
Nature Works	Cargill Dow	140 000	500 000 (in 2010)
	Hycail		50 000–100 000
Polyhydroxyalkanoates			
Biopol	Metabolix	11 000	50 000 (ADM/Metabolix joint venture in 2008)*
Nodax	Joint agreement Procter & Gamble and Kaneka	2500	50 000
Starch based			
Mater-Bi	Novamont	35 000	
Solanyl	Rodenburg	40 000	
Ecofoam	National Starch	20 000	
Cereplast	Cereplast		Approx. 20 000
Biopar	BIOP		10 000
Based on petroleum feedstock			
BAK	Bayer	4000	
Ecoflex	BASF	7000–8000**	36 000 (14 000)
EASTAR BIO	Eastman	14 000	
PCL	Union Carbide	>5000	
PVAXX	PVAXX	14 000	91 000
Total		293 500	Approx. 850 000

* In 2007 Metabolix and ADM (global company of agricultural processing) announced that they will jointly produce Mirel™ through their joint venture Tellus™. First commercial plant in Iowa (USA) is expected to start up in 2008 and will produce Mirel at an annual rate of 110 million pounds.
** BASF started up a new plant in 2006 for the production of its biodegradable plastic Ecoflex® at its Schwarzheide site in Germany, thus almost doubling its capacity for this product. The new plant will have a total capacity of 6000 metric tonnes a year, and complements the 8000 metric tonne plant in Ludwigshafen. BASF also produces a blended polymer called Ecovio, a plastic made from 45% Nature Works PLA and 55% Ecoflex.

In some of theses countries taxes on shopping bags have been introduced [9]. In March 2002, the Republic of Ireland, for example, became the first country to introduce a plastic bag tax, or PlasTax. Since 2003 a law in Taiwan requires restaurants, supermarkets and convenience stores to charge customers for plastic bags and utensils.

The German Parliament (the Bundestag) and the Assembly of the German States (the Bundesrat) have approved a regulation in the German Packaging Ordinance granting far-reaching privileges to certified compostable packaging, and thus enabling a collection system to be implemented in parallel with the increasing amounts being used. Market experts expect an almost immediate and fast-growing market for compostable packaging in Germany. All the large retail chains have prepared for the introduction of compostable packaging – ALDI's carrot bag test in late 2004 in southern Germany is only one example [10].

Following the amendment to the German Packaging Ordinance, all certified packaging (packaging that is biodegradable according to DIN EN 13432) will be exempt until 2012 from the DSD recycling fee (DSD – Duales System Deutschland), irrespective of the raw material basis. The amended German Packaging Ordinance makes special provision for certified biopackaging, i.e. packaging proven to be compostable: for a limited period during the market launch, such products need not be accepted as returns, nor are they subject to recycling quotas.

In California the two new state environmental laws are expected to provide opportunity for bio-based plastics. Assembly Bill 2147, the "truth in labeling" law, and Assembly Bill 32, the Global Warming Solutions Act, are expected to encourage the use of biodegradable and compostable plastics [11]. To provide consumers with accurate, useful information, the "truth in labeling" law mandates that environmental marketing claims for compostable plastic food or beverage containers must follow rigorous, uniform and recognized standards. Plastic containers bearing the terms "biodegradable", "compostable" and "degradable" must meet current standard specifications established by the American Society for Testing and Materials (ASTM).

Moreover, the Federal BioBased Products Preferred Procurement Program provided that Federal Agencies in the USA must give purchasing preference to bio-based products designated by this programme [9]. The authority for this programme is included in the Farm and Rural Investment Act (FSRIA) of 2002. To be designated for preferred procurement, items of single use bioplastic must meet appropriate ASTM standards for biodegradability. Some examples are cutlery, rubbish bags or food containers.

During the 2005 World Exposition in Aichi (Japan) more than 20 million biomass-derived biodegradable plastic items were used [12].

9.4. DEMAND ESTIMATION

In 1998, the demand for biodegradable polymers was estimated at about 70 000 tonnes in 2001[13], and another forecast gave a much higher value, i.e. 200 000 tonnes [14]. As can be seen from Table 9.2 the forecast was quite reasonable.

Under the European Climate Change Programme (ECCP) estimates were made for the production of bio-based polymers (and other bio-based materials) until 2010. According to these estimates bio-based polymers are expected to grow in the European Union from 25 kt in 1998 to 500 kt in 2010 without supportive policies and measures (P&Ms) and to 1000 kt with P&Ms [2]. Novamont expects that half or more than half of all bio-based produced in 2010 will be starch polymers, i.e. 250 to 500 kt.

The International Biodegradable Polymers Association and Working Groups (IBAW, Berlin) follows this view and projects a further growth of bio-based polymers in the EU to 2–4 million tonnes until 2020. Half of this total is expected to consist of compostable products while the other half would be durables.

The Japanese Biodegradable Plastics Society (BPS) has prepared projections for the market of biodegradable polymers in Japan. By 2010, the total consumption is estimated at 200 kt of which 187 are expected to be bio-based.

Bioplastics are beginning to enter automotive, computer and consumer electronic markets.

The leading companies, e.g. DuPont and British Petroleum, have recognized the importance of sustainable development and have been incorporating it as a key strategic element in operating their companies [15]. DuPont and others are working hard to create demonstrably more sustainable products, take them to market, and communicate both their value to the consumer and their sustainable advantages. DuPont intends that by 2010 the company will make 25% of its products using renewable materials [16].

The biodegradable polymer market in the US has witnessed a series of consolidations over the past ten years [17]. The early industry participants were unsuccessful due to costly manufacturing technologies and absence of enough legislation to fuel the market. The situation, however, has recently changed, with the boost for biodegradable polymers being observed in the US market. In 2003 there were more than seven players in the US biodegradable polymer market. The US biodegradable polymer market revenues were estimated at 27.0 million USD, corresponding to 22.9 million pounds of unit shipment [17]. According to Frost & Sullivan – the business research & consulting firm – the US biodegradable polymer market revenue is forecast to grow to 67.8 million USD in 2010 at a compound annual growth rate (CAGR) of 12.2%. The emerging issues of landfills and composting and the growing awareness of environmentally friendly products encourage interest in the use of compostable polymers. Apart from the traditional products such as compost bags, agricultural mulch films, leaf and lawn waste bags many industrial participants have been innovating end user applications for the new age market, including automotive plastics, compact discs, fibres, drug delivery systems, medical devices, cellular phones, computer parts, tyre fillers, cameras, and food and beverage containers. An example of new high-tech applications of compostable polymer materials is a new grade of polylactic acid polymer designed as the casing for some of Sony Walkman personal stereo products, developed jointly by Mitsubishi Plastics and Sony Corp.

In Japan, the demand for biodegradable plastics is estimated to be 200 000 tonnes in 2010 and 1.5 million tonnes in 2015 [18].

9.5. CONCLUSIONS

Major requirements for the commercial success of compostable polymer materials are: price, market demand, performance, composting infrastructures and legislation. The increased interest in compostable polymers is a response to the growing price of petroleum and growing consumer demand for sustainable products. The global demand for renewable polymers is experiencing rapid growth as costs issues have improved in recent years. Improvement in processing technologies and effect of scale-up decreased the price of compostable polymers. According to Cargill Dow [19] in 2006 demands for PLA exceeded supply.

Factors such as soaring oil prices, worldwide interest in renewable resources, growing concern regarding greenhouse emissions and a new emphasis on waste management have created renewed interest in biodegradable polymers. Improved biodegradable plastics performance has also a positive influence on global demand.

According to IBAW [20] with regard to the optimized manufacturing processes and improved cost competitiveness of the future, the long-term perspectives for bioplastics are promising.

The potential of compostable polymers materials depends on a further reduction in price, development of the cost of fossil resources, process optimization and scaling up supported by full life cycle assessment analysis, legislative measures, and development of composting infrastructures and the environmental awareness of consumers.

REFERENCES

[1] Stapert H.R. PhD Thesis, Wageningen, 1998, Environmentally degradable polyesters, poly(ether-amide)s and poly(ester-urethane)s.
[2] Techno-economic feasibility of large-scale production of bio-based polymers in Europe (PRO-BIP), Final Report, Utrecht/Karlsruhe, October 2004.
[3] Mecking S.: Nature or petrochemistry? Biologicallly degradable materials, *Angew. Chem. Int. Ed.* 43 (2004) 1078.
[4] Akiyama M., Tsuge T., Doi Y.: Environmental life cycle comparison of polyhydroxyalkanoates produced from renewable carbon resources by bacterial fermentation. *Polym. Deg. Stab.* 80 (2003) 183.
[5] Products from plants – the biorefinery future, Outputs from the EPOBIO Workshop, Wageningen, The Netherlands, 22–24 May 2006.
[6] de Bragança R.M., Fowler P.: Industrial markets for starch, http://www.nnfcc.co.uk/nnfcclibrary/reports
[7] Biodegradable inks for packaging gravure printings, Toyo Ink MFG, Ltd, October 2004.
[8] www.biopolymers.nl
[9] *Bioplastics Magazine*, 02/2006, vol. 1, p. 10.
[10] Reske J.: Beauty of bioplastics, www.earthscan.co.uk/news
[11] http://omnexus.com/news; available 12/06/2006.
[12] Report from the Government of Japan on the EU – Japan Business Dialogue Round Table (the BDRT) recommendations, March 2006.
[13] Wood A., Scott A.: *Chem.Week*, 10 June 1998.
[14] Anonim, Macplas E2 (1998) 36.
[15] Dorsch R.R., Miller R.W.: Carbon, carbon, everywhere, nor any drop to…market, *J. Ind. Ecology* 7 (2004) 13–15.
[16] *Chemical Market Reporter*, 27 August 2001.
[17] www.frost.com/prod/servlet
[18] www.nodax.com/kaneka_pgchemicals.htm
[19] O'Connor R.P.: Biomass refining for the production of PLA and ethanol, 28 September 2006, University of Minnesota, USA.
[20] Bioplastics at the leading edge of change. IBAW, Press release, 30 January 2006, Berlin.

Index